What Makes Flamingos Pink?

Other Books by Bill McLain

Do Fish Drink Water?

What Makes Flamingos Pink?

A Colorful Collection of Q & A's for the Unquenchably Curious

Bill McLain

Quill
A HarperResource Book
An Imprint of HarperCollins*Publishers*

Coca-Cola, Coke, and Fresca are registered trademarks of the Coca-Cola Company.
Excedrin is a registered trademark of Bristol-Myers Squibb Company.
Frisbee is a registered trademark of WHAM-O, Inc.
Gatorade is a registered trademark of the Quaker Oats Company.
Gerber is a registered trademark of Novartis.
Gill-line is a registered trademark of the Gill-line Company.
Midol and Vanquish are registered trademarks of the Bayer Corporation.
Oscar is a registered trademark of the Academy of Motion Picture Arts and Sciences.
Pepsi and Mountain Dew are registered trademarks of PepsiCo, Inc.
Ping-Pong is a registered trademark of Parker Brothers, Inc.
Sanka is a registered trademark of Kraft Foods Inc.
Selectric is a registered trademark of the IBM Corporation.
Silly Putty is a registered trademark of Binney & Smith, Inc.
"The day the music died" is a registered trademark of Don McLean.
Tylenol is a registered trademark of McNeil Consumer Healthcare.
Velcro is a registered trademark of Velcro Industries B.V.
Yahtzee is a registered trademark of Hasbro, Inc.

HarperCollins books may be purchased for educational, business, or sales promotional use. For information please write: Special Markets Department, HarperCollins Publishers Inc., 10 East 53rd Street, New York, NY 10022.

First HarperResource Quill paperback edition published 2002

Designed by William Ruoto

Library of Congress Cataloging-in-Publication Data has been applied for.

ISBN 0-06-019826-5
ISBN 0-06-000024-4 (pbk.)

04 05 06 RRD 20 19 18 17 16 15 14 13 12

To Snar, with love

Contents

5 FAR OUT

6 FOOD

7 THE HUMAN BODY

8 INVENTIONS

9 LANGUAGE

10 ENTERTAINMENT

11 SCIENCE

12 SPORTS and GAMES

13 TRANSPORTATION AND TRAVEL

14 UNITED STATES

15 WEATHER

16 WORLD

Acknowledgments

Although an author gets the credit for writing a book, there are always others who are instrumental in its creation. I would like to thank them.

First of all, I want to thank Jeff Simek and Karen Arena of Xerox who have encouraged and supported me from the beginning and have always been there when I needed them. I also want to thank Mark Resch of CommerceNet for his continuing support.

I want to thank Sonia Saruba, not only for her wonderful job in editing my original draft but also for her suggestions, comments, and support.

I also want to thank the people at HarperCollins, especially my editor, Greg Chaput, whose insight and recommendations greatly enhanced the book; Jay Dea for his creativity in designing the wonderful book jacket; Virginia McRae for her excellent job on editing the final draft, checking my facts, and keeping me honest. And a special thanks to all of the other people at HarperCollins who contributed their time and talents.

Finally, I want to thank the thousands of people who have sent me questions. Without them, this book could never have been written.

1

Animal Kingdom

Do all bears hibernate? (Bear with me on this one.)

To be perfectly accurate, no bear hibernates. When an animal hibernates, it is near death and may appear to be dead. Its body temperature drops to near 32°F, it breathes just a few times a minute, and its heartbeat is so slow as to be almost imperceptible. If the animal is exposed to warmth, a few hours may pass before it awakens and is alert. Animals that hibernate include bats, hedgehogs, ground squirrels, and marmots.

Although most people say that bears hibernate, hibernation is not the proper term in the case of bears, because during their sleep there is little change in their body temperature, respiration, or metabolic rate. A better term would be "deep winter sleep" (the scientific term is "dormancy"). If left alone they can sleep in the same position for months, yet they awaken quite easily if disturbed.

The reason for hibernation and deep winter sleep is the same. During winter, food is scarce, and it's often difficult for some animals to maintain their normal body temperature. To protect themselves, many animals pass the time away by hibernating or sleeping. They store body fat when food is plentiful, then live off the stored fat when hibernating or sleeping. By sleeping for long periods of time, they keep activity to a minimum and also control their temperature and metabolism so that no growth occurs.

A bear may decide to sleep in a hollow tree, a cave, a pile of brush, or a den that it has dug. Sometimes the bear adds dried leaves and grass to its bed for additional insulation against the harsh cold of winter.

Not all bears sleep all winter. If bears live in an environment with a good year-round food supply, they don't need to go into a deep winter sleep. Tropical bears such as sun bears, sloth bears, and spectacled bears never go into a winter sleep.

The male polar bear never goes into dormancy, but the female does only if she is pregnant.

If a bear is accustomed to dormancy but is kept in a zoo where food is always available, it will not go into dormancy regardless of how cold it may get.

FACTOIDS

The Australian koala bear is not a bear at all but a marsupial related to the kangaroo. The bearcat, a nickname for the Southeast Asian binturongs, is not a bear either. They are related to a little-known group of animals that includes civets, genets, and linsangs.

People seem to love bears in spite of their size and ferocity, especially the fictional bears Baloo, Fozzie Bear, Paddington Bear, Yogi Bear, and Winnie the Pooh. The most famous live bear was Smokey.

Polar bears have white fur but black skin. Each hair is actually a clear hollow tube designed to funnel the sun's rays to the bear's skin, thereby keeping it warm. Because the rays bounce off the fur, the polar bear appears to be white.

The sloth bear lives on a diet of termites. However, like humans, bears have a sweet tooth. They often break open beehives and will continue eating honey even though their nose has been stung many times.

When a bear is dormant, it does not eliminate its waste but recycles it by turning the toxic compounds into protein. Researchers are trying to discover how bears do this because the bears' method could lead to methods of treating kidney failure in humans.

DID YOU KNOW?

Although bears are meat eaters and often ferocious, they have their gentle side too. In 1995 four unwanted kittens were dumped near a wildlife rehabilitation center in Grant's Pass, Oregon. Although employees trapped three of the kittens and took care of them, they could not catch the fourth.

By late summer the kitten was starving. Seeing a 560-pound

grizzly bear devouring food in the compound, the kitten squeezed through a hole in the fence and approached the grizzly.

Everyone was terrified that the poor stray kitten was going to be the grizzly's next meal. The bear looked at the kitten, pulled a piece of chicken from its dinner, and tossed it aside for the kitten.

The bear, named Griz, never harmed the kitten. In fact they became close friends and ate, slept, and played together. The employees named the kitten "Cat."

At last report, Griz and Cat were still the best of friends. They probably still are today.

Is it true that a dinosaur larger than *Tyrannosaurus rex* was recently found? (A head-to-head match.)

In 1995 scientists in Argentina discovered a meat-eating dinosaur they named *Giganotosaurus*. They claimed that it was as big as or bigger than the North American *Tyrannosaurus*. However, their claim is still being disputed. Although *Giganotosaurus* had a larger skull, it had a smaller brain, making it less intelligent. It has a longer upper leg bone but a shorter lower leg bone, so both dinosaurs were about the same height.

However, the non-meat-eating dinosaurs were considerably larger. In fact, of all the known dinosaurs, which one is the biggest depends on how you measure a dinosaur's size.

A dinosaur named *Argentinosaurus* was recently discovered in Argentina (hence the name). It was 70 feet high, 120 feet long, and weighed around 220,000 pounds. In other words, it was as tall as a seven-story building, almost the width of a football field, and weighed around 110 tons.

In 1994 scientists in southeastern Oklahoma found the bones of another huge dinosaur. They named it *Sauroposeidon,* which means "earthquake god lizard." It was 60 feet high, weighed 60 tons, and was 150 feet long, partly because it had the longest neck of any known dinosaur.

One standard measurement used to define a dinosaur's size is length, which is measured from the tip of its nose to the tip of its tail. Because *Sauroposeidon* was 30 feet longer than *Argentinosaurus,* some scientists claim it is the largest animal ever found.

However, other experts claim that *Argentinosaurus* was the largest dinosaur ever found because it was 10 feet taller and weighed almost twice as much.

We don't want to be unpatriotic, nor do we have anything against the state of Oklahoma, but our vote for the largest dinosaur on record goes to the monster from Argentina.

That's not the end of the story. A fossil hunter in Colorado unearthed one of the largest dinosaur legs ever discovered. Based on the leg, he estimated that the dinosaur was probably 98 feet long and weighed 130 tons. Unfortunately, that's only a guess until someone finds the rest of the dinosaur. But the fossil hunter has already given it a name: *Ultrasaurus,* or the "ultimate" dinosaur.

FACTOIDS

The longest dinosaur was 150 feet long, while the shortest was only 20 inches long, about the size of a modern chicken.

The word *dinosaur* was coined by Sir Richard Owen. It means "fearfully great lizard," from the Greek *deinos*, meaning "fearfully great," and *sauros*, meaning "lizard."

Some dinosaurs are named for their features (head, tail, claw, teeth, or feet), some are named for a person, some are named for the place where they were found, and some for their behavior. For instance, the *Albertosaurus* was discovered in Alberta, Canada; the *Lambeosaurus* was named after Lawrence Lambe; and the term "velociraptor" means "speedy robber."

Dinosaurs could neither swim nor fly but were all land animals. The flying pterosaurs and swimming ichthyosaurs were closely related but were not true dinosaurs.

Most experts today, following a theory first formulated in the late 1800s by the biologist Thomas Huxley, believe that birds are technically dinosaurs.

DID YOU KNOW?

Many people think that dinosaurs became extinct about 65 million years ago. That's not quite true. Prior to that time, dinosaur families were already dwindling, and a number of species had already become extinct.

However, 65 million years ago some catastrophe occurred that not only destroyed all land animals weighing more than about 55 pounds but also killed many smaller life forms. All of the dinosaurs were obliterated, as well as some marsupials, fish, snails, sea urchins, bird families, and over half the plankton groups.

The theory as to what caused this catastrophe most widely accepted today is that an asteroid 5 to 10 miles in diameter hit the earth, penetrated its crust, scattered dust and debris into the atmosphere, and caused severe storms and volcanic eruptions. The dust and debris thrown into the atmosphere blocked out all sunlight for months. The atmosphere itself was changed and had higher concentrations of sulfuric and nitric acid.

Because of acid rain and lack of sun, plants died out with two effects: a depletion of oxygen levels suffocated many smaller organisms, and plant-eating creatures soon starved. The larger meat-eating animals no longer had any prey once the plant-eating animals died, so they began eating each other and eventually died out.

We often use the word "dinosaur" to mean something unwieldy and inefficient. However, dinosaurs were highly efficient for their size. They became extinct because something from another world crashed into earth and changed the environment.

We pride ourselves on being the most efficient of all creatures. Yet we too could become extinct if another deadly visitor from space crashed into our planet.

Do spiders and other insects ever sleep? (Are they dream spinners?)

A spider is not an insect, or a "bug." Bugs have six legs and three body parts while a spider has eight legs and only two body parts. A spider is an arachnid, a family of air-breathing invertebrates that includes scorpions and ticks.

Scientists are still arguing whether true insects can sleep. Some argue that because insects don't have eyelids, they can't shut their eyes to sleep. They claim that only mammals sleep and that it's silly to talk about sleeping insects or fish. They hope that by defining sleep as something only mammals do, they won't have to worry about how something without eyelids can sleep.

However, other scientists who study bugs and arachnids have noted that these creatures display daily periods of inactivity that must be called sleep.

A moth will become very inactive in the daytime and will even tuck its antennae under its wings. You can gently poke it and it won't move. It's a very sound sleeper. Honey bees rest at night in a manner much like a deep sleep. Other insects display similar behavior.

Most scientists now agree that insects and spiders do actually sleep. It's a good thing that having eyelids isn't necessary for sleep because a typical spider has eight eyes.

FACTOIDS

The innocent daddy longlegs spider is more venomous than a black widow spider, but it can't open its jaws wide enough to bite a human.

The silk that a spider spins for its web is about five times stronger than steel.

Every state in the United States has a "state insect." Butterflies are the state insects of 20 states and honey bees the insect of 16 states. The other 14 state insects range from a ladybug to a praying mantis.

There are about 35,000 different kinds of spiders in the world.

Spiders come in all sizes. The female goliath tarantula of South America can reach a leg span of 10 inches, about the diameter of a dinner plate. The male Patu digua spider's body is smaller than a pinhead.

NASA scientists studied the effects of marijuana, Benzedrine, caffeine, and sleeping pills on spiders. Spiders on marijuana tried spinning webs but gave up about halfway through. Those on Benzedrine ("speed") spun webs very quickly but left large holes in them, creating weird abstract patterns. Those on caffeine could only spin a few random strands while those on sleeping pills never got started at all. Who says spiders can't sleep?

DID YOU KNOW?

A dew-covered spider web glistening in the early morning sunlight is a beautiful sight. Yet not all spiders spin webs, and some spin silk for reasons other than to catch prey.

Some spiders, such as the *Segestria,* live in a hole in the ground and line the walls with silk to make the home more comfortable to live in. This spider spins trip lines out of the hole for a short distance. A trip line vibrates if an insect touches it, and the spider will rush out of its hole and grab the insect.

The common house spider also uses trip lines, but instead of just a few, the spider creates a messy sheet of silk in front of its door.

The net-casting spider makes a net of silk web and then drops it on any unsuspecting prey unfortunate enough to pass under it.

The bolas spider emits a chemical substance that mimics the sex attractant of certain female moths. If a male moth is attracted to the sex scent, the spider swings a strand of silk with a sticky blob on the end to catch the poor male and haul him in.

All young spiders engage in ballooning, by which they release a long silken thread and then float on the wind to move to a new area.

Spiders that spin webs must be fast, because insects can escape quickly. If not attacked by the spider, a fly can escape a web in about five seconds.

A spider must also be cautious, because many things can be caught in its web that not only destroy the web but also hurt the spider. Grasshoppers, crickets, bees, and wasps are not welcome visitors to the spider's parlor.

There is an old superstition that if you can't find your cows, hold a daddy longlegs spider by his back legs, and one of his front legs will point in the direction where your cows are. If you should ever lose a cow and want a daddy longlegs spider to help you find it, don't bother the poor spider if he's sleeping.

What is the fastest snake in the world? (It can't outrun a Dodge Viper.)

The fastest snake in the world is the black mamba, which can reach speeds as fast as 10 to 12 miles per hour in short bursts over the ground. It has been known to chase people.

The black mamba is not only fast, agile, and ferocious, it is also one of the deadliest snakes in the world. Just two drops of its venom can kill you. Even a minor scratch can prove fatal. If bitten, a victim will usually die within four hours or less. Until an antivenin was developed in the 1960s, the bite of a black mamba was 100 percent fatal. It often takes as many as 10 vials of antivenin to save a victim.

Running into a black mamba is hardly a pleasant experience. This slender snake is usually about 9 feet long, but it's not uncommon to find one 10 to 12 feet long. It's a nervous creature and often will not let a human approach within 75 feet or so. However, it angers easily and if annoyed will raise its head and front body as much as four to six feet off the ground. It shakes its head from side to side while giving out a long and very frightening hiss.

The black mamba never bluffs and is quick to attack. It will speed past you to get away while at the same time biting you numerous times. No part of your body is safe because the black mamba's head might be at your eye level when it raises its body.

Fortunately for us, there are no black mambas in the United States. You only have to worry if you go to parts of Africa such as Botswana, Namibia, Zimbabwe, Angola, Zaire, and parts of South Africa. However, a snake that is just about as deadly as the black mamba is Australia's taipan snake. Maybe Dorothy had it right when she said, "There's no place like home."

FACTOIDS

Hoop snakes do not bite their tails and roll downhill like a wheel when they are frightened. This is a myth. It probably started because they often lie in a coiled position resembling a hoop.

Although the king cobra rarely bites humans, it has enough venom to kill an elephant.

Of all the snakes in the world, only about 10 percent are poisonous. Yet many people kill a snake on sight without bothering to determine if it's harmless or not.

The state of Hawaii has no snakes at all. In fact, there is a $25,000 fine for importing a snake of any type. Snakes like to eat birds' eggs, and Hawaii is known for its great variety of colorful birds. Around 1950, the brown tree snake was accidentally brought into Guam, and since then the snakes have virtually wiped out the native forest birds of Guam. Twelve species of birds, some found nowhere else, have disappeared from the island. Hawaiian officials are fearful that the brown tree snake might slither into their state. A few were found years ago, but they were all dead. Nonetheless, Hawaii is keeping a careful eye out for the brown invader.

DID YOU KNOW?

Whether snakes are revered or hated depends greatly on the culture. Everyone in the Judeo-Christian culture has read about how the serpent in the garden of Eden enticed Eve to eat the forbidden fruit. We also admire Saint Patrick for getting rid of all the snakes in Ireland.

In other cultures the snake is not necessarily hated. In fact, it is often a powerful religious symbol. According to Greek legend, Aesculapius discovered medicine by watching a snake use herbs to bring a dead companion back to life. Even today, the symbol of a doctor or medicine consists of two snakes wrapped around a staff.

A major Aztec god was Quetzalcoatl, the "plumed serpent." Aztec legends say that the god told them to build their city in the spot where they saw an eagle with a serpent in its mouth. The only problem was that when they saw the eagle holding a serpent, it was perched on a small island in the middle of a lake.

The Aztecs were undaunted. They founded their city where they had seen the eagle. They built floating homes and barges filled with dirt so they could grow crops. They reclaimed the land until all of the smaller islands formed one large island. Eventually, the lake was drained completely.

Today we know this city as Mexico City, one of the most populous in the world. And it all started because an eagle snatched up a snake.

Are zebras white with black stripes or black with white stripes? (Of course a zorse is not a horse.)

Yes. That's the answer. A zebra can be white with black stripes or it can be black with white stripes. In fact, there are zebras with dark brown stripes and zebras that are either all white or all black.

Scientists believe there are two reasons for a zebra's stripes. The first is camouflage. Zebra stripes break up the outline of its body, making it difficult for a predator to identify it. When a zebra is alone, moving among the tall grasses of the plains, it looks just like wind-blown grass and is difficult for a predator to see. A group of zebras will all huddle together, making it difficult for a predator to single out one zebra amid the mass of moving stripes.

The second reason is to cool the zebra in the hot African sun. Zebras have a shiny coat that can dissipate up to 70 percent of the sun's heat. The black stripes can get hotter than the white stripes by as much as 50°F.

Although there are eight types of zebras in the world, the three most well-known types are in Africa. Each type has a different style of stripes. The Grévy's zebra is considered the most beautiful, because it has very thin and closely spaced stripes. The plains zebra has very wide vertical stripes that bend to become horizontal on the zebra's rump. The mountain zebras have broad black stripes that extend down the legs but do not cover the off-white belly.

Every zebra in the world has a unique pattern of stripes. There are no two alike, just as no two human fingerprints are alike. Zebras recognize each other by looking at the pattern of stripes.

Although a zebra's stripes make it a strikingly beautiful animal, they are not there for decoration but are designed to provide camouflage, cooling, and identification.

FACTOIDS

When plains zebras are frightened, they gather together in a cluster so that the myriad moving stripes confuse predators. On the other hand, Grévy's zebras primarily use speed and stamina to escape predators. They can run as fast as 40 mph over reasonably long distances.

Virtually all attempts to domesticate zebras so they can be ridden or used as draft animals have failed.

When zebras are first born, they are brown and white. Within an hour after birth, the newborn can run as fast as the herd and can recognize its mother by sight.

When a herd of zebras is sleeping, one always stays awake and alert for predators. In effect, it is doing sentry duty.

If a family member becomes separated or lost for some reason, the other zebras in the family will search for it. The family will also adjust the speed of its travel so that the old and weak can keep up with the group.

Zebras in captivity have been successfully mated with other equine species. Of course, the names always begin with the letter *z*. A zorse is the offspring of a zebra stallion and a horse mare. A zonkey, sometimes called a zebrass, or a zedonk, is the offspring of a zebra stallion and a female donkey.

DID YOU KNOW?

Zebras are not the only creatures that use camouflage. A white polar bear blends in with the snow, while animals that live in the desert are often yellowish or tan so that they blend in with the desert environment.

However, snow, desert, forest, and mountains are not striped. So why do some animals, such as the zebra, have stripes? For years scientists have argued about the reason for striped animals. Today there are two prevalent theories.

The more accepted theory suggests that a striped or spotted animal mimics light filtering through the trees or high grass and thus is camouflaged.

Another theory suggests that a black-and-white color scheme is an effective warning device, much like stripes on a railroad crossing gate. Stripes are not unique to mammals but are also found on birds, reptiles, and insects. Stripes on a polecat, a marbled salamander, a pied kingfisher, and a carabid beetle warn predators that the striped creature is dangerous or inedible. A biologist did an experiment in which cats and hornets were allowed to taste the flesh of 38 different types of birds. The "least edible" rating was given to the only three birds that were black and white.

Although scientists still do not completely agree on why some animals are striped, one fact they all agree on: a zebra is a strikingly beautiful animal.

Why don't squirrels fall when they run across a telephone wire? (This is not a squirrely answer.)

Squirrels can run across wires because they are agile, have good coordination, and most of all because they have an excellent sense of balance. They also use their large bushy tails as a "balancing pole," which helps.

A squirrel has five toes on each foot. The sharp claw on each toe is slightly curved, which lets the squirrel wrap its paws around objects with small diameters, such as a wire, rope, or small branch. However, a squirrel has a big problem with anything that's over an inch in diameter. For these larger objects, the squirrel must use its claws as spikes. That's why it's so easy for a squirrel to climb a tree. On the other hand, it's almost impossible for a squirrel to climb a large object made of glass or metal.

A squirrel's tail is not used just as a balancing pole. If a squirrel falls, its tail serves as a parachute. Squirrels can fall more than 100 feet without hurting themselves. The versatile tail keeps the squirrel dry during rainstorms, and warm during cold nights, and shades the squirrel from the sun on hot days. A squirrel also uses its tail as a flag to warn other squirrels of imminent danger. If a squirrel ends up fighting, it uses its tail as a shield.

Another versatile part of a squirrel is its eyes. Because of their position, a squirrel can see above, below, and behind without moving its head. However, it has trouble seeing straight ahead. That is the reason you'll see a squirrel move its head from side to side before jumping.

Rather than saying "squirrely" or "nutty as a squirrel," it might be better to say "versatile."

FACTOIDS

Squirrels come in all sizes. The African pygmy squirrel is only five inches long, while the giant Asian squirrels are over three feet long and can jump twenty to thirty feet.

There are over 300 species of squirrels found in every country in the world except Madagascar, Australia, and the polar regions.

Not all United States presidents have felt the same way about squirrels. Teddy Roosevelt kept pet flying squirrels in the White House. Ronald Reagan gathered acorns at Camp David and fed them to the squirrels when he returned to the White House. He even commissioned an artist to paint a picture of a squirrel running across the White House lawn. However, Dwight Eisenhower hated squirrels and didn't want them on his private golf course. He had them trapped and shipped elsewhere.

Only two places in the United States have large populations of white squirrels: Marionville, Missouri, and Olney, Illinois.

In 1791 the land surrounding the town of Belpre, Ohio, was invaded by squirrels, which ate the entire corn crop. They then swam across the Ohio River and started devouring crops in West Virginia.

DID YOU KNOW?

One of the most fascinating species of squirrels is the flying squirrel. They don't actually fly like a bird; rather, they glide.

The squirrel first climbs high in a tree and then moves its head back and forth to check out its flight path. It propels itself from the tree and extends all four legs to stretch out a membrane that connects the front paws to the rear paws. This membrane serves as a sail so the squirrel can glide through the air, often as far as 50 yards or more. This is quite far when you consider that the typical flying squirrel is only three or four inches long and weighs only a few ounces.

While gliding, the squirrel can steer around obstacles by using its versatile tail as a rudder. It usually lands rear feet first on the trunk of another tree and immediately scurries to the opposite side of the tree in case a predator is pursuing it. Because of this gliding ability, it's possible for a squirrel to sail the length of a football field if it starts out from a perch 100 feet in the air.

In the United States the southern flying squirrel can be found in the southern and eastern parts of the country, while the northern flying squirrel is found throughout California, Idaho, Montana, the Great Lakes region, and the Appalachian mountains. They live in deciduous forests.

You might not have seen a flying squirrel, because they are nocturnal creatures. If you want to see one, autumn nights are best, because the squirrels are busy gathering food for winter.

If you don't have a forest nearby, you might be lucky enough to have an attic in your house. Flying squirrels have been known to nest in attics.

What makes flamingos pink? (It's not sunburn.)

The color of most birds is a matter of genetics. Bluebirds are blue by nature, robins have red breasts, and cardinals are bright red. The flamingo is different: flamingos are pink because of what they eat.

Depending on the species, the color of a flamingo can range from pale pink to crimson. A flamingo eats extremely small water plants such as algae and tiny animals such as insect larvae and crustaceans. Most of this food is full of natural pigments called carotenoids. The pigment in the carotenoids makes the flamingo turn pink or red as new feathers grow.

When flamingos are in captivity, such as in a zoo, they are fed a natural red coloring agent called Roxanthin or they are given a carotenoid-rich diet of prawns, shrimps, and crustaceans so that they retain their pink color. If flamingos are given food that lacks carotenoids, the pink color will eventually fade away.

There are five species of flamingos: the greater flamingo (which includes the American or Caribbean flamingo), the lesser flamingo, Chilean, Andean, and the James's flamingo. Although the different species range in color from pink to red, the color is still completely dependent on the diet. For example, lesser flamingos eat only carotenoid-rich algae, called spirulina, and thus have a more intense color than the greater flamingos, which get the carotenoids secondhand by eating creatures that have already digested the algae.

When we eat carrots or beets we are also eating carotenoids, just like the flamingos. We just can't eat enough beets and carrots to make us change color. The only way we can turn pink is either to blush or get a sunburn.

FACTOIDS

Although it appears that a flamingo's knees bend backward, they do not. The middle leg joint that appears to be a knee is actually an ankle. A flamingo's knees are higher up and hidden beneath its feathers. Its long legs bend the same way as ours but its body parts are in different proportions.

Ancient Romans considered flamingo tongues to be a great delicacy.

The only species of flamingo with yellow legs and feet is the Andean flamingo. It also has a red spot between its nostrils.

A flamingo eats with its head upside down. It scoops up water and mud and uses its large tongue to pump the solution through a filter in its beak that catches the food. A flamingo can filter up to 20 beakfuls of water in just one second.

Flamingos always walk on their toes.

Whether standing in water or on land, a flamingo can stand for hours on one leg, with the other leg folded under its belly and the head laid along its back.

Flamingos are very noisy. They make sounds such as nasal honking, grunting, and growling.

A flamingo can live to be over 50 years old.

DID YOU KNOW?

Many flamingos live around hot volcanic lakes. The very name flamingo is associated with fire and brings to mind the mythical firebird, the phoenix.

In both Greek and Egyptian mythology, the phoenix was a bird as large as an eagle with bright scarlet and gold feathers and a melodious cry. Every morning the phoenix would bathe in the cool water of a nearby well and sing a song so beautiful that the sun god would stop his chariot to listen.

There was only one phoenix in the world at any given time and it lived for at least 500 years. When the phoenix felt death coming, it would build a nest of aromatic twigs, set it on fire, and be consumed by the flames. A new phoenix would then miraculously emerge from the funeral nest. In one version of the legend, it would embalm its ancestors's ashes in an egg made of myrrh, then fly to the City of the Sun (Heliopolis) to lay the egg on the altar of the sun god Ra.

Egyptians associated the phoenix with immortality, an idea that carried on through Roman times. The Romans thought that Rome would last forever. In fact, they called it the Eternal City. The image of a phoenix appeared on Roman coins as a symbol of undying Rome.

In 1965 Jimmy Stewart appeared in a wonderful film called *The Flight of the Phoenix*. In the film a cargo plane crashes in the Libyan desert during a fierce sandstorm. Although the twelve men survive the crash, they know that search planes will not find them, and they are short of food and water.

One of the men, a designer of model airplanes, proposes that they rip off the one undamaged wing and use it as the basis for an airplane that they will build to fly them to an oasis 200 miles away.

They succeed, and the new plane emerges from the wreckage of the old, just as the phoenix arises from its own ashes.

Some say that the phoenix represents the ability to leave the world and its problems behind while flying toward the sun in clean, clear blue skies. Not a bad idea at all.

More questions? Try these Web sites.

Animal Bytes

http://www.selu.com/bio/wildlife/links/animals.html

This page lists mammals, birds, reptiles, amphibians, fish, insects, and invertebrates. Just click on the one you're interested in to see a list of fast facts, a list of fun facts, and information about ecology and conservation. For example, you'll discover that a dromedary camel can live to be 50 years old and it can drink 30 gallons of water in just 10 minutes.

Pets also Need a Vacation

http://ask.4anything.com/4/0,1001,611,00.html ?%3B029125i

This site contains a great deal of information about traveling with your pet. It has links to four sites that will help you plan a vacation with your pet: Animal Adventures, Romping Good Time, Travel Supplies, and Relocating Rover.

Squirrels

http://www.amy-from-mars.com/links02.html

Just about everything you ever wanted to know about squirrels. A partial list of the contents includes care and feeding, health and welfare, species, squirrels in the news, outwitting squirrels, squirrel clubs, flying squirrels, and links to other squirrel sites.

Animal Planet

http://animal.discovery.com/animal.html

This is the Discovery Channel's Web site and is loaded with information about all types of animals, including the latest animal news. "Earth alert" lets you track animal happenings around the world, and "Animal guide" not only covers everything from crocodiles to wolves but also has links to other animal Web sites.

Dinosaur Links
http://dinosaur.umbc.edu/links/index.html

This Web site lists links to a number of sites with information about dinosaurs. It also tells you how to get on a dinosaur mailing list, how to obtain commercial dinosaur items, and even lists where scientists are currently digging for dinosaur fossils.

This page also has a gallery of dinosaur art and lists a number of sites especially designed for children.

2

Crime

How many innocent people have been executed by law in the United States since 1900? (Justice is often blind.)

Although we work hard to provide justice to all, sometimes it doesn't work the way it should. According to a 1997 article in the *Salt Lake City Weekly,* researchers have reported that in the twentieth century, 139 innocent people were sentenced to death and of these, 23 were executed. One study concluded that if the number of innocent victims executed is compared to the number of prisoners on death row, then one out of every 100 prisoners executed will be innocent. Another more sobering study stated that one out of every 7 executed prisoners is innocent.

Unfortunately, the convicted were found to be innocent not through normal legal channels but as the result of investigations by journalists, the application of new scientific techniques, or the efforts of college students or the dedicated lawyers who were convinced their clients were innocent.

Criminals who are poor rarely have expert attorneys. There are many instances of low-paid lawyers who do not prepare for the trial, fail to interview witnesses, and actually sleep through a murder trial. As terrible as this sounds, courts often justify it. As one Texas judge said publicly, "The Constitution says that everyone's entitled to an attorney of their choice. But the Constitution does not say that the lawyer has to be awake."

There are other reasons why people are wrongfully convicted. When a high-profile crime occurs, such as the murder of a police officer or the brutal slaying of a child, the police are subjected to extreme pressure to solve the crime as quickly as possible. At other times there is no eyewitness to the murder and prosecutors often rely on less reliable sources. The media also plays a role in convictions. Publicity that may be inadmissible or misleading can influence jurors to ignore legitimate doubts about a suspect's innocence. There are also "death qualified" juries in which jurors who do not believe in the death penalty are removed from the juror pool.

The arguments for and against the death penalty will probably rage on for years to come. But whether the death penalty is in place or not, there is an obligation to do everything possible to keep innocent victims from being convicted of crimes.

FACTOIDS

In recent years, the state of Texas has executed far more prisoners than any other state. The number of executions seems to vary greatly by geographical region. For example, in the past twenty-five years, Texas and Virginia alone accounted for over 55 percent of all executions in the South, where 481 prisoners were executed. The Midwest executed 64 prisoners and the West executed 50. The Northeast had only 3 executions during that time.

In the past decade almost twice as many death row inmates have been found innocent when compared with the previous 20 years.

Most states use lethal injections or electrocution for executions. Five states use the gas chamber, while two use hanging and two use a firing squad. States using the gas chamber, hanging, and the firing squad use lethal injection as an alternative method.

Thirty-eight states have the death penalty, while twelve states and the District of Columbia do not.

One man on death row was released just two days before his execution date. It turned out that law officers suppressed statements from reliable witnesses who saw the victim alive hours after the alleged murder. A key witness also admitted to lying.

DID YOU KNOW?

You might have seen the television series or movie *The Fugitive,* which tells the story of Dr. Richard Kimble, accused of murdering his wife. Kimble claimed he had been knocked unconscious by an intruder with a limp who was the murderer.

The story of *The Fugitive* was based on a factual case, one of the most sensational trials of the 20th century. Dr. Sam Sheppard, a prominent Cleveland osteopath, was accused of murdering his wife. Although he claimed to be innocent and said the murderer was an intruder with a limp, he was convicted of second-degree murder and sentenced to life imprisonment.

After serving 10 years in prison, Sheppard was released when an appeals court said he did not have a fair trial, and two years later the Supreme Court granted him a new trial. This time the jury acquitted him, 12 years after he had been arrested for murder. He died of liver failure four years later at age 46.

Although the controversy about his guilt or innocence continued after his death, his son worked to clear his father's name. Sam Sheppard's body was exhumed in 1998, and DNA tests provided convincing evidence that he was wrongly convicted of killing his wife.

It took 44 years for Dr. Sam Sheppard to be exonerated.

What was the largest bank robbery in U.S. history? (Watch out for revolving doors.)

It was February 10, 1997, and the Lakewood, Washington, branch of the Seafirst bank had just closed. A well-dressed man in a trench coat and FBI cap used some excuse to get the teller to open the door. Once inside he put on sunglasses, pulled out a gun, and let in an accomplice, also wearing sunglasses and a trench coat.

They were extremely polite as they ushered the three women tellers into the vault and bound their hands with plastic handcuffs. The men then began filling bags with money and taking the bags out to their car. Most of the bills they took were hundreds, fifties, and twenties. The loot probably weighed around 100 pounds because they made off with $4.5 million dollars. A few hours later the tellers managed to free themselves and call 911, but by then the men had vanished.

The two, William Kirkpatrick of Hovland, Minnesota, and

Ray Bowman of Kansas City, Missouri, might have gotten away with it if they had been as smart at handling money as they were at robbing banks.

Although his share was over $2 million, Bowman forgot to pay the $30 rent on his storage locker. When the manager opened up the storage space, he found guns, silencers, police radio scanners, and materials for creating disguises. He quickly called the police.

Just one month after the robbery, Kirkpatrick was stopped for speeding by a Nebraska state trooper. When the trooper searched the trunk of Kirkpatrick's car, he found guns and $1.8 million in cash.

If Bowman had paid his storage bill and Kirkpatrick had driven just a little bit slower, they might have been successful in getting away with the largest bank robbery in U.S. history.

FACTOIDS

Bowman and Kirkpatrick had robbed other banks, and the FBI said that the two "trench coat bandits" had stolen a total of almost $8 million. Neither man ever filed a tax return.

In over 75 percent of bank robberies, no gun is shown, just a note. Even customers in the same line often don't realize the person in front of them has just robbed the bank.

Hawaii's first bank robbery occurred in 1934 when two bandits stole $976 from a bank teller. One man had used an eyebrow pencil to paint a large Groucho Marx mustache on his face. They were caught within hours, and there wasn't another bank robbery in Hawaii until 1955, over twenty years later.

Nearly 50 percent of all bank robberies take place on Friday.

DID YOU KNOW?

Had Bowman and Kirkpatrick not been so careless, they might not have been caught. However, many bank robbers are simply inept. Here are a few examples.

Three men planning to rob a bank in Scotland got stuck in the revolving door. After being helped by the bank staff, they left, only to return shortly and tell everyone they were robbing the bank. When they demanded cash the head teller laughed, thinking it was a joke. One man jumped over the counter to get at the money, but fell and broke his leg. The other two men tried to escape but got stuck in the revolving door again and were caught and arrested.

In Durham, North Carolina, a would-be robber broke through a glass window and climbed down a rope to get into the bank. Unfortunately, he slipped and hurt himself. He then realized that the money vault was locked and he had no way of getting out of the bank. The robber called 911 for help.

In Maryland one bank had its night deposit box on the sidewalk. Some enterprising robbers had the brilliant idea of hooking a chain to it and using their 4-wheel-drive vehicle to pull it loose from the concrete. It was a good theory, but when they threw the vehicle in gear, it lunged forward and ripped off the rear bumper. Frightened by the noise, the robbers took off. They didn't get far. When police arrived at the scene, the bumper was still chained to the night deposit box, with the license plate securely attached.

Two bank robbers in France spent hours drilling into a bank vault from an attached building. Their calculations were slightly off. After hours of hard work they broke through—into the bathroom.

One man successfully robbed a bank. He wasn't caught until the next day when he went back to the same bank to deposit the stolen money into his account.

Jesse James, Bonnie and Clyde, and other notorious bank robbers might not have been very nice people, but at least they weren't stupid.

What are the differences among first-degree, second-degree, and third-degree murder? (It's not always a matter of degree.)

Murders are handled by police homicide detectives, so let's begin by defining homicide. Homicide is not murder. It is a general term meaning that someone caused the death of another human being either directly or indirectly. It may or may not be a crime.

Some types of homicide are criminal offenses such as murder, manslaughter, and criminal negligence. Other types of homicide are not crimes, such as self-defense or accidental death.

The term murder means that a person kills another person, or intends to kill another person, or intends to do bodily harm that will probably cause death. There are only two degrees of murder, first and second. There is no such thing as third-degree murder.

A first-degree murder means that the murder was deliberately planned, or as we often hear, was premeditated. If a murder was not premeditated, it can still be first-degree murder if a peace officer or jail guard is killed in the line of duty or if the murder occurred during certain crimes such as kidnapping or sexual assault.

The definition of second-degree murder is easy. It's any murder that is not a first-degree murder.

Manslaughter is a homicide without malicious intent or premeditation. It can be voluntary or involuntary. State laws differ on occasion, but voluntary manslaughter is considered a crime because the death results from failure to safeguard human life. If a drunk driver kills someone, for example, it is voluntary manslaughter.

Involuntary manslaughter means that a person is killed while someone is committing a misdemeanor or by some type of negligence. It can also be killing in self-defense. It is usually not considered a felony.

FACTOIDS

About half of all murder victims knew their killer and about 20 percent of all murders involve family members. Of all women murdered, about one in four are killed by a husband or boyfriend.

Although serial killers capture news headlines and intrigue the public, only one murder out of 100 is committed by a serial killer. The chances that you will be killed by a serial killer are extremely slim; your chances of being killed by a family member are much higher.

DID YOU KNOW?

One of the United States's most famous murder trials has virtually been forgotten. It happened in 1907 when the former governor of Idaho, Frank Steunenberg, was assassinated.

The Western Federation of Miners had been warring with mine owners for years. The miners had put 60 boxes of dynamite underneath the world's largest concentrator and blew it to bits. Steunenberg, who was governor at the time, asked President McKinley to send in federal troops, which he did. Every male in every union-controlled town was arrested, even doctors and ministers. They were transported by boxcars and jailed in an old barn. Over 1,000 men were held prisoner without a trial. This did not go over well with the miners, who were suspected of hatching the plot to assassinate the governor.

The prime suspect was Harry Orchard who, although he initially insisted he was innocent, later decided to turn state's evidence. Orchard admitted that he had killed the former governor and 17 other people as well. He also said that he had tried to kill the governor of Colorado, two Colorado Supreme Court justices, and the president of a mining company, but had failed in all attempts. He claimed that the murders had been ordered by labor leader William "Big Bill" Haywood.

Haywood was thus accused of ordering the assassination as well as other bombings and murders over a period of 15 years.

The famous attorney Clarence Darrow defended Haywood. His defense is considered by many to be the best of his career. He convinced the jurors that convicting Haywood was an attack on organized labor. The jury deliberated just under eight hours and returned a "not guilty" verdict.

That was not the end of Haywood's troubles. In 1918 he was convicted of espionage and sedition for inciting a strike in a war-sensitive industry and was sentenced to 30 years in prison. While out on bail waiting for an appeal, he fled to the Soviet Union, where he became allied with the Bolsheviks. He died in Moscow in 1928.

On the other hand, Harry Orchard, at the time America's most famous mass murderer, spent the rest of his life in prison.

Is it true that there was a man who loved his girlfriend so much that when she died he dug up her grave and kept her corpse in his home? (Not until death do us part.)

Karl von Cosel, a German X-ray technician, had fallen hopelessly in love with a beautiful Cuban patient, Elena Hoyos Mesa, who was being treated for tuberculosis at a Key West, Florida, hospital. Elena was married but her husband had deserted her when she became ill. She loved Karl and said she would marry him once she became divorced. However, it was not meant to be; she died first.

Death did not stop Karl from being a faithful lover. He built a large mausoleum for Elena and visited her daily. One night, while visiting Elena, he was startled by a noise, opened the girl's outer casket, and later claimed that Elena started talking to him. From that time forward he conversed with her every night. Eventually, he took her body out of the casket, cleaned it, reconstructed it with wax and plaster, dressed it, and kept it propped up in his bed. Karl believed that he could resurrect Elena and then fly away with her in his airplane.

Key West was a small town in 1940, and rumors that Elena was not in her casket reached her sister, Nana. To prove that he was treating Elena well, Karl took Nana to his home, showed her Elena's body, and claimed that she was still alive.

Karl was arrested shortly thereafter. However, he was not convicted because of the statue of limitations. While he was in jail, the authorities buried Elena's body in secret so he couldn't find it. Because Karl had lost his job and was broke, after he got out of jail he charged tourists who wanted to see his home and sold souvenirs of Elena to raise money.

Because of his notoriety, Karl left Key West and moved to Zephyrhills, Florida, where he sold postcards of the dead girl.

FACTOIDS

The practice of firing three rifle shots over a grave during a funeral is based on the old custom of stopping the fighting to remove the dead from the battlefield. Once the dead had been cared for, three volleys were fired to signal continuation of the battle.

When Elena's body was moved to a funeral home, over 5,000 people came to see it. Some suggested that she be put in a glass case like Sleeping Beauty and displayed as a tourist attraction.

Pygmies in the African Congo are not comfortable with death. When a tribe member dies, the tribe pulls down the hut on top of the deceased person. While relatives cry, the rest of the tribe move their camp to a new location. The dead person is never mentioned again.

It is not uncommon for a corpse to be left for the animals. Solomon Islanders used to lay their dead on a reef for the sharks to eat, and the Parsees of Bombay, India, used to put their dead on top of towers to be eaten by vultures.

DID YOU KNOW?

When we think of famous burials, we don't often think of a horse, but perhaps we should recall Ruffian, the racehorse who refused to be beaten.

Ruffian was a filly that won her very first race by 13 lengths and broke the track record. In her first two years of racing she won every race she entered. Late in her second year of racing she fractured a bone and was retired to rest for a while. Even so, she was crowned two-year-old filly of the year.

As a three-year-old, she raced longer distances and won every time. She won all three legs of the Filly Triple Crown and seemed destined for greatness. But there was one more test.

It was decided to hold a match race between Ruffian and the winner of that year's Kentucky Derby, Foolish Pleasure. Over 31,000 people watched the match.

Foolish Pleasure took an early lead but Ruffian came on strong and pulled in front by half a length. As they reached the mile marker, the two great horses were running neck and neck.

Then it happened. There was a sickening crack. Ruffian had broken her leg.

Her jockey struggled to pull her up but she was fighting him. She wanted to continue the race in spite of her shattered leg.

A team of four veterinarians and an orthopedic surgeon worked for 12 hours trying to save her leg, during which time she stopped breathing twice and had to be revived. When the anesthetic wore off, she was disoriented, and although they tried to restrain her, she flailed about and broke her other leg. The veterinarians knew she could not survive further surgery so they put the gallant horse to sleep to end her suffering.

Ruffian was buried at Belmont Park, near the flagpole. As a fitting tribute, they buried her with her nose facing the finish line.

Flags flew at half mast that sad day.

Who was the famous bandit who wore a metal bucket on his head? (It wasn't for fashion.)

In the 1800s, outlaws in the rural parts of Australia were known as "bushrangers": after they had robbed a stagecoach or bank, they would head for the bush to escape. The last, and probably most famous, bushranger was Ned Kelly. Some people considered him a criminal; others considered him to be a folk hero, much like Billy the Kid and Jesse James in the United States.

The government in rural Australia in those days was badly administered, and many people believed that it was corrupt. Wealthy landowners, called "squatters," had all the best land. People who were poor, such as Ned Kelly's family, could buy land from the government. They were called selectors. The government required the selectors to improve the land, but the soil was so poor the families couldn't make a living. In order to survive, the selectors had to resort to stealing.

Ned Kelly was only 15 years old when he was arrested for stealing a chicken, but was found not guilty. Three weeks later, he was arrested for having a stolen horse in his possession. Although he had no idea that the horse had been stolen, he was sentenced to three years of hard labor. When he was finally released from prison, he discovered that the local police had stolen all but one of his 32 horses. In spite of feeling cheated and persecuted, he tried to stay on the right side of the law, for a while. When police arrested Ned's mother, he feared that he might be next, so he and his brother decided to hide in the bush, and the Kelly gang was eventually formed.

The Kelly gang began staging a series of daring robberies. Ned wore a handmade suit of armor to protect him from police bullets. The helmet was fashioned from a metal bucket and looked bizarre at best. Because they only robbed from the rich, they soon captured the public's imagination. People felt that they were striking out at a corrupt system that was suppressing the poor.

When the gang ran across three policemen, Ned thought they meant to kill him and in the ensuing gun battle, he killed all three. He was able to escape because the Kelly gang had become folk heroes and many people helped them elude the police.

Ned made his last stand in 1880, when police surrounded the hotel where he was staying. Rather than try to escape, he decided to fight the police, believing his armor would protect him.

In a blazing gun battle, Ned Kelly's gang fought off 30 policemen for seven hours. Although Ned looked comical in his homemade suit of armor, the police bullets just ricocheted off him. Finally, a police officer realized that Ned's legs were unprotected and starting shooting at his legs. Once he was wounded, the police captured him.

All of the Kelly gang except Ned were killed in the gun battle. Ned's armor saved him from the bullets but not from the gallows. He was hanged on November 11, 1880. He was only 25 years old.

FACTOIDS

Ned Kelly's complete suit of armor weighed 97 pounds.

Whenever he robbed someone, Ned gave the person a letter explaining to the government how the police had persecuted him.

The reward for the capture of the Kelly gang eventually reached 8,000 pounds, which today would be almost $2 million. The reward didn't help much.

Just before Ned Kelly was hanged, 60,000 people signed a petition seeking mercy for him.

After being captured, Ned Kelly said, "If my lips teach the public that men are made mad by bad treatment, and if the police are taught that they may exasperate to madness men they persecute and ill treat, my life will not be entirely thrown away."

DID YOU KNOW?

Until around 1850, most bushrangers were escaped convicts. After that they were mostly settlers who had run into trouble with the law. Although some were ruthless killers, others treated their victims humanely. In fact one of them, Edward Davis, shared his loot with the poor. But whether they were humane or ruthless, they usually ended their career at the end of a hangman's rope. The bushrangers disappeared after 1880 when the last one, Ned Kelly, was hanged.

Although Ned Kelly has been dead for over 120 years, Australians still argue about the truth of the legend. Some see him as a hardened killer who deserved to be executed. Others see him as the victim of a corrupt system which forced him into crime. They think he was a martyr.

What is the real truth? Probably somewhere between the two, but no one knows for certain.

Who was Dr. H. H. Holmes? (Elementary, my dear Watson.)

In 1896 no one could have imagined that the "Monster of 63rd Street" and his "castle of horror" would ever be forgotten. Yet today few people know his name or have any idea that he murdered more than 200 people, one by one.

The mass murderer's real name was Herman Webster Mudgett. While attending medical school in Michigan he stole corpses, used acid to disfigure them so they couldn't be identified, and then collected on the life insurance policies he had bought using fictitious names. When he was finally caught stealing a corpse, he ran away, moved to Chicago, and changed his name to Dr. H. H. Holmes.

In Chicago he opened a respectable pharmacy and eventually built a drugstore empire that made him a fortune. In 1890 he bought a vacant lot and started construction of a magnificent mansion that had 100 rooms. However, it was not a typical mansion. It had a maze of secret passages, trap doors, chutes, fake walls, acid vats, and secret entrances.

In 1893 Chicago was filled with tourists who were taking in the great Chicago Exposition. The timing was perfect for Holmes. His mansion had just been completed and he began renting rooms to the tourists. Very few ever returned to visit the Chicago Exposition. He killed most of his lodgers, using the same insurance fraud scheme he had used in the past.

He also lured young women into his "castle," promising to marry them. Instead, after forcing them to sign over their life savings, he drugged them and threw them into one of the empty shafts in the house. When a victim awoke, she found herself trapped behind a glass panel and struggled vainly as Holmes pumped lethal gas into the chamber. The unfeeling Holmes shoved the body down a chute to the basement, where he had acid and lime vats waiting to disfigure the corpse.

When police began to be suspicious of Holmes's activities, he set fire to the mansion and left Chicago. He continued his grisly activities in Missouri, Pennsylvania, and Texas. Texas was his undoing. He made the mistake of stealing a horse, which was a capital offense in Texas. After being arrested as a horse thief, police searched his burned-out mansion and found the remains of over 200 corpses.

He was hanged in 1896.

FACTOIDS

Murder was not the only illegal activity of H. H. Holmes. Once he flavored city water with vanilla and sold it as a cure-all called Linden Grove Mineral Water. Another time he used his credit to buy a very large safe, which he moved into his mansion. He built a room around the safe. When he refused to pay for the safe, his creditors came to take it back, but they could not get it out of the tiny room.

Holmes's record of 200 murders was only broken 80 years later when Pedro Alonso Lopez, a native of Colombia, murdered 300 people in a bloody orgy that covered three countries. Known as the "Monster of the Andes," he was discovered when a flash flood uncovered some of his early victims.

DID YOU KNOW?

The doors that opened into brick walls and the stairs that led nowhere in the mansion of H. H. Holmes bring to mind another bizarre mansion, the Winchester Mystery House in San Jose, California.

Sarah Winchester was the wife of William Winchester, who made a fortune as the owner of the Winchester Repeating Arms Company, manufacturer of the famous Winchester repeating rifle. When William died of tuberculosis, Sarah was instantly wealthy with an income of over $1,000 a day, but the money meant nothing to her. She grieved so much at her husband's death that she sought solace from a medium. The medium told her that the family was cursed because the firearms produced by the company had killed so many thousands of people. She told Sarah to move west and build a home for herself and the spirits of those who died as the result of the Winchester rifles.

Sara bought a 17-room house in San Jose, California, and started building, and building, and building. The medium had said that if Sarah stopped construction, something terrible would happen to her. She kept an army of workmen busy day and night for 38 years. Eventually the house grew into a 160-room Victorian mansion seven stories high. It was a modern marvel at the time, with heating and sewer systems, pushbutton-operated gas lights, three elevators, and 47 fireplaces.

It had numerous staircases that went nowhere, a chimney that stopped inside the house, doors that opened to blank walls, trap doors, and doors that opened straight out onto the lawn, several stories below.

Sarah also had an obsession with the number 13. Most of the windows had 13 panes of glass, walls had 13 panels, there were 13 cupolas, and all but one staircase had 13 steps.

In 1922 Sarah Winchester died in her sleep at the age of 83. All construction on the house ended the next day.

More questions? Try these web sites.

CRIME LIBRARY
http://www.crimelibrary.com/

This Web site has five major sections, each containing a great deal of information.

"Classic crime stories" includes detailed stories about the Black Dahlia, Lizzie Borden, the Borgias, and Leopold and Loeb.

"Serial murders" starts out with a profile of serial killers to help you understand what makes them tick and then includes stories about famous serial killers, including the Son of Sam, Jack the Ripper, and the Hillside Strangler.

"G-Men" talks about famous lawmen such as Wyatt Earp and Eliot Ness, as well as famous criminals such as Al Capone, John Dillinger, Baby Face Nelson, and Murder Incorporated.

"Spies" has stories of famous spies and assassins such as John Wilkes Booth, the Cambridge spies, and Carlos the Jackal.

The site also has "Crime news," which is updated daily, covering major crimes occurring around the world.

WINCHESTER MYSTERY HOUSE
http://www.winchestermysteryhouse.com/

The official site of the Winchester Mystery House in San Jose, California. It includes the history of the house, a list of amazing facts, and a QuickTime movie you can view. It also lists special events and tours.

FEDERAL BUREAU OF INVESTIGATION
http://www.fbi.gov/

This is the official site of the FBI and lists the most wanted criminals, current major investigations, and crime reports. It also tells you how to apply for a job with the bureau and what the benefits are.

Forensics

http://library.thinkquest.org/17049/gather/

Click on "Reference" and you'll see a number of topics that you can check out such as ballistics, DNA, hair and fibers, fingerprints, and so on. These aren't covered in detail but give you a quick overview of the topic.

Are You a Potential Victim of Crime?

http://www.Nashville.Net/%7Epolice/risk/

This is the site of the Nashville Police Department in Tennessee. If you click on "Rate your risk," you can take various tests to see how likely you are to be mugged, murdered, or buglarized and what you can do to reduce your risk.

If you click on "Safety tips," you'll see a list of topics you can select for more information, such as: "Learn some handy self-protection tips," "Sprays? Shockers? Alarms? Which protective device is right for you?"; and "What do you do if you're a victim of domestic abuse?"

The Crime of the Century

http://www.fbi.gov/yourfbi/history/famcases/brinks/brinks.htm

This is the story of the famous Brinks robbery in 1950. A group of armed men broke into a Brinks office, tied up the employees, and made off with over $1.2 million in cash and over $1.55 million in checks, money orders, and other securities. It took the FBI six years to solve the crime.

3

Customs

How did the custom of kissing originate? (This isn't about the chocolate kind.)

One theory of kissing prevalent among anthropologists contends that kissing originated from "premastication." Although today we can buy soft baby food, this is not true in many cultures and especially in ancient primitive cultures. Premastication basically means "pre-chewing." A mother chews the food and then pushes the soft, pre-chewed food into her infant's mouth. This process involves mouth-to-mouth contact and is sometimes called "kiss-feeding." Many anthropologists think that is how kissing originated.

Another popular theory, and a less messy one, asserts that kissing originated from a custom symbolizing the union of souls. In many cultures, individuals put their faces together to symbolize a spiritual union. Because they believed that breath was part of a person's soul, by exchanging breaths, they were intermingling their souls. Some authorities believe that this practice eventually led to the Inuit custom of rubbing noses together as a sign of affection and love, as well as to our custom of kissing.

Our culture recognizes many types of kisses. There is the kiss between parent and child, which represents love and affection. There are social kisses, which might be a quick kiss to the cheek when friends meet and occasionally when meeting someone for the first time. The ceremonial kiss, as the kissing on both cheeks when heads of state meet, is not a social convention but a political symbol signifying the good intentions of both parties.

There is also the romantic kiss between lovers or spouses. There's no need to explain that to anyone.

FACTOIDS

In the 1941 film, *You're in the Army Now*, Jane Wyman and Regis Toomey kissed for three minutes and five seconds, the longest kiss in film history.

The practice of kissing virtually died out during the bubonic plague in the 1600s, because people were afraid of spreading the plague by direct contact with one another. Fortunately, once the plague ended, kissing came back into style.

It is believed that over 5 million women have kissed the armor-clad statue of Italian soldier Guidarello Guidarelli in Ravenna, Italy, because they believe that so doing will guarantee a happy marriage.

In the 1926 film *Don Juan*, the hero kissed a variety of

señoritas 191 times, or an average of one kiss every 53 seconds. Seven seconds more and he could have been a "Minute Man."

The first kiss ever recorded on film occurred in 1896, when John C. Rice kissed May Irwin in a film called, naturally, *The Kiss.*

When someone in the military or industry uses the word kiss, they very often are using an acronym for "Keep It Simple, Stupid!"

DID YOU KNOW?

We must enjoy kissing, because it always seems to be cropping up in our language and our music. We even eat chocolate "kisses."

There are hundreds of "kiss" phrases such as "kiss and make up," "kiss and tell," and "kiss of death."

Song titles are filled with the word: "Kiss to Build a Dream On," "Hold Me, Thrill Me, Kiss Me," "Kisses Sweeter Than Wine," "Then He Kissed Me," "Sealed with a Kiss," and many, many more. There is even a rock group called "Kiss."

We even kiss rocks! In the small village of Blarney, Ireland, on the top story of a 90-foot high castle, is the famous Blarney Stone. Legend says that if you kiss the stone you will receive the gift of eloquence. Kissing the stone is not that easy, however. In order to reach the stone to kiss it, you have to lean over backward and downward while someone holds your feet.

The legend of the Blarney Stone says that an old woman cast a spell over the stone as a reward for a king who had saved her from drowning. When he kissed the stone, the king could speak sweetly and convincingly.

Another folk legend claims that many years ago the ruler of the castle was told he had to give his fortress to Queen Elizabeth I to prove his loyalty. He said he would be more than happy to do that. However, whenever he was about to give up the castle, it seemed that at the last minute he had some excuse to prevent doing so. His excuses became frequent but quite plausible, and became a joke in the royal court. When the eloquent excuses of the castle owner were relayed to the queen, she replied, "Odds bodkins, more Blarney talk!" Today we use the term "blarney" to mean "an ability to influence without giving offense."

And that's no blarney!

Why is a wedding ring worn on the third finger of the left hand? (A path to the heart, for eternity.)

As far as we know, the ancient Egyptians were the first to place a ring on the third finger of the left hand to signify the marriage union. It was placed on that finger because Egyptians believed that the "vein of love" ran from this finger to the heart. They used a ring because they believed that the circle was the symbol for eternity. It represented perfection because it had no beginning and no end.

Rings found in ancient Egyptian tombs were made of pure gold. The name or title of the owner was engraved on the ring in hieroglyphs. The poorer citizens of Egypt wore rings of silver, bronze, amber, ivory, or simply glazed pottery.

Because gold was precious to the early Romans, a gold ring symbolized everlasting love and commitment.

King Edward VI of England decreed that the third finger on the left hand was to be the ring finger. In the 1549 *Book of Common Prayer,* the left hand was designated as the marriage hand.

From the earliest times in our history, people have always given advice to newly married couples such as "comfort each other," "respect one another," and "listen to each other." One of my personal favorites is "Never yell at each other unless the house is on fire!"

FACTOIDS

The old wedding phrase "Something old, something new, something borrowed, something blue" has a definite history. Something "old" referred to a personal gift from the bride's mother to provide a bond to the bride's old life and family. Something "new" signified hope for the future and was symbolic of the new family to be formed by the married couple. Something "borrowed" was to be a gift from a happily married woman. The gift was supposed to carry some of the married woman's happiness into the new marriage. Something "blue" had two different meanings. Ancient Roman maidens wore blue because it denoted modesty and fidelity, while for Christians the blue is associated with the purity of the Virgin Mary.

The origin of the wedding shower is based on the legend of a Dutch maiden who fell in love with a poor miller. Her family could not afford a dowry so their friends "showered" them with gifts so that they could be married without a dowry.

Hindus tie old shoes on vehicles leaving the wedding ceremony as a sign of good luck.

Because an Anglo-Saxon bride was often kidnapped before a wedding, she stood to the left of the groom so his sword hand would be free. The best warrior in the tribe stood next to the groom to help him defend his bride. That is why in today's weddings, the best man stands to the right of the groom.

Many cultures believe that loud noises scare away evil spirits. Today the tradition continues with our custom of the bridal party honking their horns when leaving the wedding.

In medieval times, Europeans believed that newly married couples were very vulnerable to evil spirits. If the groom carried the bride, she was protected from the floor and the evil spirits in the ground. That is the origin of the custom of carrying the bride over the threshold.

DID YOU KNOW?

Our wedding customs and traditions come from diverse ethnic and cultural backgrounds. Although most of us are familiar with our own wedding customs, there are many fascinating customs in other times and countries.

Because African-American slaves were not permitted to marry, they publicly declared their love by jumping over a broom to symbolize jumping through a doorway from single life to the domestic life. Couples had their hands bound together at the wrists so that they were symbolically linked. Although some people believe this is the origin of the phrase "tying the knot," many other cultures have a similar ceremony. For example, in a Hindu marriage ceremony, the bridegroom hangs a ribbon on the bride's neck and then ties it in a knot. The ancient Carthaginians bound the thumbs of the betrothed with leather thongs.

In small Italian villages, the newlyweds walk to the town plaza, where there is a sawhorse, a log, and a double-handled saw. With the crowd cheering them on, they must saw the log apart. This symbolizes that in all of life's trials and tribulations, the couple must always work together.

In the simple Moravian wedding ceremony, the bride and groom together light one large candle. Every guest has a handmade beeswax candle. One guest lights his candle from the large candle and then uses his candle to light the candle of the guest next to him. This continues until everyone in the church has a lit candle, symbolic of the warmth of love from family and friends.

No matter what the country, a wedding is filled with love.

How long has the running of the bulls been held and how many people have died? (Don't mess with these dangerous creatures, and that's no bull.)

The running of the bulls, known as the *encierro,* has a very long history, although, the current event in Pamplona, Spain, began around 1852 as part of the nine-day festival of San Fermin, usually in July.

The tradition started when it was necessary to drive the bulls from their corral to the bullring. At first, drovers prodded the bulls along the way, but eventually the boys of the butchers guild, who were responsible for buying the bulls, began running in front of the bulls to show their courage. They felt if they didn't run and prove their bravery, no Pamplona girl would ever marry them. That's how the tradition started.

Since the running of the bulls started, 13 people have been killed, all since 1924. Prior to 1924, there is no record of anyone being killed, mainly because no one kept any records. In that year a young man received a mortal stab wound to his lungs just as the bulls were entering the ring. In 1995 Peter Matthews Tasio died on the horns of a bull, and the photograph was seen all over the world.

In 1947 a bull named Semillero killed two people. In 1980 the bull Antioquio also killed two people, one at the town hall and another in the bullring. In 1997 the last person to die was a Spanish onlooker who fell from a high stone wall.

In addition to the deaths, there have been a number of near-fatal injuries as well as a large number of contusions. An American was lifted by a bull's horns and shaken furiously before being tossed to the ground. A Swede was lifted the same way and flipped around for over 10 seconds before escaping.

Pamplona, Spain, is not the only place that promotes the running of the bulls. The United States now has its own version of the famous Pamplona event. It started in 1998 and is held annually in the small town of Mesquite, Nevada, near the Arizona border. Runners pay $50 each to be chased by twenty 1,500-pound Mexican fighting bulls down a narrow course as long as six football fields. There were no fatalities in the inaugural event, although a few people were injured.

FACTOIDS

The running of the bulls occurs every morning during the nine-day fiesta. It's estimated that more than 2,000 runners take part.

The run itself lasts, only a couple of minutes. The excitement is in the intensity of the spectacle and the risk of injury or death.

A fighting bull is as fast as a race horse for a short distance, can turn sharper than a polo pony, can hook a falling leaf, can overturn an automobile with its tossing muscle, is considered one of the smartest animals, and is the only animal known to instinctively attack a man or a horse.

DID YOU KNOW?

If you should ever decide to partake in the running of the bulls, here are some tips for you.

Never enter the fray if you've been drinking. It's a good way to get killed.

Keep your eye on the bull (not the ball). Bulls are very fast and can catch up to you very quickly. You also need to watch the runners in front of you. If one of them trips and you don't notice, you'll fall on him and the bull is going to be on top of both of you.

Never try to get the bull's attention. The bull simply wants to get out of the tight path as quickly as possible and will go straight to the bullring if you let him. If you get his attention and he gets separated from the group, you'll have to confront him, which might be terribly unpleasant.

If you fall, stay down. The bull will just step on you as he passes, as will the other bulls. Although you may get bruises and broken bones, it's better than standing up and getting gored by a charging bull. People have died this way in the past. Hug the ground and cover your head with your hands. A few prayers would also help.

When you get into the ring, fan out and head for the closest barrier. This will give the drovers room to drive the bulls into the waiting pens. If you go to the center of the ring, you'll be in the path of the bulls—not a good spot to be in.

Here's the best tip of all: stay home and watch the running of the bulls on television.

Why is there a crescent moon on outhouse doors? (A symbol of the times.)

The main reason for carving anything into an outhouse door is for light. Outhouses didn't have electric lights and putting in a window wasn't a very good idea. So builders of outhouses carved a design into the door, above the line of sight, to let in light. The carved design let in sufficient light during the day, and at night the moon would also shine through the cutout. The hole also helped with ventilation.

Although any design could be carved, in most pictures a crescent moon is carved in the outhouse. Most authorities believe that the use of this symbol dates back to the 1500s or 1600s.

Because people were largely illiterate at that time, putting up signs saying MEN and WOMEN wouldn't have been much help. So the builders used cutouts of the moon and sun to let people know which outhouse to use. The moon has long been accepted as representing women, while the sun represented men. These symbols also helped foreign travelers who might spend the night at an inn. It didn't matter what language they spoke because the symbols told them which outhouse to use, much like the universal nonverbal symbols we use today for "no smoking," "telephone," "slippery when wet," and so on.

If one of the outhouses at an inn was damaged or destroyed and could no longer be used, it was automatically assumed to be the men's outhouse. The reasoning was that men could always go behind a tree, so the crescent moon was put on the remaining usable structure for use by women. For economy, many inns only constructed an outhouse for women. This custom soon became so widespread that eventually the moon became a symbol used for all outhouses.

FACTOIDS

Most outhouses had two different-size holes, a larger one for adults and a smaller one for children.

In a federal park in Pennsylvania, park rangers wanted people to have a nice rest room facility, so they designed a two-hole outhouse without running water. The park service spared no expense to create a unique structure. It was built of quarried limestone and had a gabled roof. It cost $333,000 and took two years to build, being completed in 1996. It's even earthquake proof.

An Illinois woman publishes an outhouse newsletter. She has also donated most of her 700 outhouse knickknacks to a Canadian museum.

Centuries ago, plumbing pipe was made from wood or earthenware. Eventually lead was used, and it took skilled workers to fit or repair pipes in a building. The Latin word for "lead" is *plumbum* from which we get the word plumber.

Many cities and counties hold annual outhouse races. Volunteers run along city streets pulling along outhouses of every design and color imaginable. Races are held in Alaska, New Mexico, New York, Texas, and Washington, to name just a few states.

DID YOU KNOW?

For some Americans, the only place to have a few minutes of quiet and solitude is their automobiles. In crowded Japan, a common sanctuary is the rest room.

An American tourist in Japan might find using some of the more innovative Japanese toilets extremely intimidating. First of all, there is the array of buttons on the toilet's keypad. The characters are in Japanese.

There are controls for raising, lowering, and warming the seat, a tiny sprayer that cleanses you with water, a temperature control for the water spray, a fan for blow-drying you after the water, and even a deodorizing fan. Oh, yes, there is also a control to flush the toilet.

Unfortunately, there are stories of tourists who never manage to comprehend the workings of these high-tech toilets. Some panic and get stuck in the stall. Others cannot figure out how to turn off the water spray and don't know how to get up without being drenched in water. There was also the case of four old high-tech toilets that caught fire. That can be very disconcerting at best.

Although the Japanese are noted for technology, if you've ever been befuddled by a high-tech Japanese toilet, you may decide that the old two-hole outhouse isn't that bad after all.

Why are barns always painted red? (Why paint them at all?)

Early barns in this country were not painted. Farmers considered painting a barn to be extravagant and showy; they simply couldn't afford it. By the mid 1800s red paint became cheap. It was made with iron oxide, which we call rust. This inexpensive paint appealed to the thrifty farmers of New England and New York. It soon became stylish to paint a barn red, a nice contrast to the typically white farmhouse. As people started migrating to the Midwest, they brought their customs with them, and it was rare to find a barn in the area that was not red.

There are other theories, however. One theory claims that barns were painted red because of the influence of Scandinavians, who painted their barns red to simulate brick. Brick buildings were a sign of wealth. Another theory suggests they were painted red to complement the green fields.

Perhaps one of the most interesting theories is a circular explanation. It asserts that red paint was readily available to farmers and was cheap because paint manufacturers made so much of it. The reason the paint companies made so much of it was that that's what the farmers wanted.

Today there are many different colors of barns. In parts of Kentucky, for example, barns are painted completely black. This is believed to be due to the tradition of creating a cheap wood preservative by using a mixture of lamp black and diesel fuel. In Pennsylvania, central Maryland, Ohio, Indiana, Illinois, and Iowa, barns are typically painted white. One guess is that when barns were used for dairy production, they were painted white to associate them with the cleanliness of their milk production. However, that's only a guess.

In other areas of the country, barns may be painted yellow, green, blue, or gray.

However, a barn painted any color other than red is still a barn.

FACTOIDS

Aside from the traditional barns, some round barns were built, often with a silo in the center, the idea being that it was more efficient to have all cows facing a central feeding area. Round barns became very popular at one time, and octagonal barns were also built as a variation.

New England farmers often built a covered walkway between the farmhouse and the barn so they could care for their cattle without having to face the elements. They also banked manure around the sides of the barn and house to keep themselves and their cows warm in the winter.

Some farmers decorate barn roofs by using different-colored roof shingles to create a pattern of a horseshoe, a cloverleaf, or even the farmer's name.

Early barns had neither ventilators for fresh air nor windows to let in light. As rooftop ventilators became more popular, farmers started enclosing them in wooden cupolas, often with elaborate designs crowned by a fanciful weathervane.

DID YOU KNOW?

People may think of George Washington as the father of our country, the commander in chief of the Revolutionary army, a great general, or our first president. All of these are true, but first and foremost, George Washington was a farmer.

Washington had an 8,000-acre plantation that he divided into five farms. He studied books on agriculture, corresponded with leading farmers around the world, and made many innovations. He was one of the first farmers to devise a system of selective breeding to raise better livestock, use crop rotation to preserve his fields, and emphasize the production of wheat and grains rather than tobacco to make the United States the world's granary.

One of his most fascinating inventions was a 16-sided treading barn. The traditional method of threshing wheat was either to beat it by hand until the grain was loosened from the straw or have horses trample on it to break the wheat from the chaff.

Washington's treading barn had two stories. The floor of the top story had half-inch gaps between the floorboards. Horses would run along a lane in the center section of the top story, trampling the wheat. The grain fell through the cracks in the floor but the chaff did not. Workers on the first floor would gather up the grain and store it for later transport to the gristmill where it was ground into flour. Not only was this a more efficient way of threshing wheat, it also protected the horses from the weather.

Most people agree that George Washington was one of our country's greatest leaders. Few people know that he was also one of the great leaders in developing our country's agriculture, which is the envy of the world today.

Why is the shamrock associated with Saint Patrick? (Where did all the snakes go?)

Although Saint Patrick is the patron saint of Ireland, he was born in Scotland in A.D. 387. He was kidnapped when he was 16 years old, taken to Ireland, and sold as a slave. While a slave, he learned to speak the Irish language fluently. After six years in captivity, he escaped to England and eventually went to Saint Martin's monastery in Tours, France, where he studied to become a priest. Saint Patrick was ordained a priest and later became a bishop. Pope Celestine I sent him back to Ireland to preach and convert the druids, an ancient Celtic priesthood, to Christianity.

Upon returning to Ireland, Saint Patrick preached to the druids in the open air. One day he was trying to explain the doctrine of the trinity. To illustrate how three were as one, he reached down to the grass growing at his feet, plucked a shamrock, and used it as an analogy of the trinity. He said it was a symbol of how the Father, the Son, and the Holy Spirit could all exist as separate parts of the same entity. He has been associated with shamrocks ever since.

Saint Patrick preached throughout Ireland and built many churches. He converted druid warrior chiefs and princes as well as thousands of their subjects. His efforts resulted in entire kingdoms converting to Christianity. He continued his preaching for 40 years. It is believed that he converted the entire country to Christianity.

Saint Patrick died in A.D. 461, at the age of 74. Although more than one town claims he was buried there, no one knows for sure where he was buried.

Saint Patrick is perhaps best known through the legend of his having driven all the snakes out of Ireland. However, most scientists believe that there never were any snakes in Ireland from the time it separated from the European continent at the end of the last Ice Age. Yet the legend is so well known, there must have been some truth to it. There is, but it's an analogy. Snakes were a metaphor for the druids. When the druids became Christians, they were no longer snakes, so there were no more snakes in Ireland.

FACTOIDS

Half the U.S. presidents have been of Irish descent.

Authorities were not sure if Saint Patrick died on March 8 or March 9. To settle the dispute, they added the two together and decreed that he died on March 17, which today is known as Saint Patrick's Day.

Shamrock comes from the Irish word *seamróg*, which means "little clover."

The parents of the notorious outlaw Billy the Kid were Irish immigrants.

Because clover plants produce three leaves, finding one with four leaves is extremely rare. That is why finding a four-leaf clover is considered good luck. However, there are growers today who have used selective propagation to produce plants with a large number of four-leaf clovers. Picked by hand, the four-leaf clovers are sold as good luck charms.

Many cultures consider the number three to be a mystical symbol, and the shamrock was often considered a mystical symbol also. In Iran it was considered to be a sacred emblem of the Persian triad. It was also considered a sacred plant among the druids and Celts.

James Hoban, who designed the White House, was an Irish immigrant.

Saint Patrick's real name was Maewyn Succat. When he became a priest, he took Patrick as his Christian name.

DID YOU KNOW?

If someone were to ask you, "What is the national symbol of Ireland?" there's a good chance your answer would be "the shamrock."Actually, it's a harp. For the past several hundred years the harp has been the symbol of the Irish and is displayed on flags, coins, and even on the Royal British coat of arms.

The Irish harp, usually referred to by its Gaelic name *clarsach,* is a small harp anywhere from three to four feet tall with 28 to 34 strings.

In medieval Ireland, a person who played the harp was called a harper. Harpers served as counselors to the kings and were given titles and wealth for their services. The king always consulted with his harper before going to war, and a harper often led the troops into battle, holding his harp in one arm and his sword in the other.

In the 1500s, probably because of fear of the harpers' status and wealth, the English Crown began hassling Irish harpers. Many were imprisoned or executed. Queen Elizabeth proclaimed, "Hang harpers, wherever found, and destroy their instruments."

Fortunately for everyone, the harps outlived the queen and are thriving in Ireland today. If you want to know how popular the Irish harp is, use any major search engine and search on "Irish harp." You'll find over 2,000 entries for recordings and music.

The *clarsach* is more than a symbol, it's part of the musical soul of Ireland.

More questions? Try these Web sites.

Wedding Customs

http://ultimatewedding.com/custom/

This site has a wealth of information about weddings, including Jewish, military, and Renaissance weddings.

On the left side of the page you can click on topics such as a song library, traditions and customs, bridal shower guide, vows and ceremonies, and articles on planning a wedding. There are also links to other wedding sites. The site also has an "Ask the experts" topic.

Outhouses

http://lest-we-forget.com/The_Outhouse/outhouse_links.htm

This page lists a number of links to information about outhouses, both factual and humorous. You can find information on building an outhouse, outhouse gifts, a tour of outhouses in the United States, and information on various outhouse races.

Native American Customs

http://www.cbtl.org/na/customs.htm

This is an excellent site about Native American customs. It covers many aspects of Native American culture, lore, and mythology and also has information on Native American clothing, totems, and pow wows, and numerous links to Native American language sites.

Everyday Eating Customs in China

http://www.cuisinenet.com/glossary/chinaday.html

Chinese cuisine has become very popular in the United States and many people have learned how to use chopsticks (if you don't know how, just click on "Use of chopsticks" for illustrated instructions). However, the customs for eating in China are different from those in the United States. This site explains Chinese eating customs and etiquette.

101 Links About Saint Patrick's Day

http://www.demanddigital.com/Saint_Patrick's_Day.htm

This site covers just about anything you could possibly want to know about Saint Patrick and Saint Patrick's Day. It has links to 101 sites, including customs and history, free online greeting cards, who the real Saint Patrick was, fun sites for children, and free screen savers.

4

Disasters

Which ship disaster resulted in the greatest loss of life? (Does it give you that sinking feeling?)

Because the sinking of the *Titanic* has generated a great deal of publicity, many people have come to believe that it was the worst ship disaster of all time. It wasn't. The greatest loss of life on a single ship occurred when the German liner the *Wilhelm Gustloff* was torpedoed by a Russian submarine in January 1945. The *Wilhelm Gustloff* was a converted luxury liner serving as a hospital ship. On the fateful night it was sunk, it was crammed with refugees, mostly women and children, and about 1,600 military personnel. Although there is no surviving record of the actual number of people on board, the most widely accepted estimate claims there were 8,600, of whom 7,700 were killed. There were only 903 survivors.

Prior to World War II, the *Wilhelm Gustloff* was the flagship of Germany's fleet of passenger liners. It had been docked for four years when the commander of the German submarine fleet ordered the evacuation of submarine personnel because of the approach of the Russian army. In January 1945, people started boarding the ship. With the German army collapsing, submarine personnel, injured soldiers, and refugees eagerly clambered aboard to escape the advancing Russians. The loading was chaotic and many parents were separated from their children. Life preservers were available for only slightly more than half of the passengers.

It was −14°F when the ship left the harbor, and layers of ice began covering the deck.

The ship was just 28 miles from the Baltic town of Leba when the Russian torpedoes hit. People panicked and many plunged overboard into the icy water. Other passengers were trampled to death in the ensuing panic. Most of the ice-covered lifeboats couldn't be lowered and of those that could, some were snagged by the bowline and spilled screaming people into the water. Other passengers chose suicide and shot themselves rather than suffer death through drowning. Thousands clung to the ship screaming for help as it sank faster and faster until it slipped beneath the waves. Then the screaming stopped and there was a deathly silence.

The greatest single ship disaster of all time was over in less than an hour.

FACTOIDS

Although approximately 1,500 people died when the *Titanic* sank, not many people know that in addition to the *Wilhelm Gustloff,* there were 19 other ship disasters that resulted in a greater loss of life than the *Titanic.*

The *Goya,* a converted German passenger ship, was hit by Russian torpedoes just before midnight on April 16, 1945. The ship broke in half almost immediately. Its masts splintered and fell, crushing passengers on the decks. Icy water rushed into the holds, drowning everyone in them. The *Goya* sank in just four minutes, with the deaths of 6,200 people.

A converted German passenger liner, the *General von Steuben,* was also sunk by a Russian submarine in 1945. Crammed with wounded soldiers and refugees, it sank in just seven minutes. Between 2,700 and 3,500 passengers died.

The *Cap Arcona* was another converted German passenger ship, loaded with inmates from evacuated concentration camps. In 1945 it was sunk by a British fighter bomber while moored in the harbor. Between 5,000 and 7,000 people lost their lives.

These three German ships alone account for the deaths of between 13,900 and 16,700 people.

DID YOU KNOW?

Military personnel in every country know that they may face death when in a combat zone. Unfortunately, many innocent people are also killed, often by their own countrymen. The sinking of ships is no exception.

In 1944, the Japanese steamer *Junyo Maru* was torpedoed and sunk by a British submarine. The British had no way of knowing that the ship carried Dutch prisoners of war and Indonesian slave laborers.

The British also sank the German steamer *Thielbek,* not knowing that many of its passengers were from a concentration camp. There were no survivors.

The Japanese steamer *Arisan Maru* was sunk by an American submarine. The submarine's crew did not know that the ship was carrying 1,800 American prisoners of war.

Perhaps one of the greatest atrocities of any war is the unnecessary and often deliberate killing of innocent civilian refugees, helpless wounded men, and prisoners of war.

What does SOS stand for? (This is going to be a lot of trouble.)

Although you may have heard that the letters SOS stand for "Save Our Souls," "Save Our Ship," or "Send Out Succor," none of these is true. The letters stand for absolutely nothing.

Samuel F. B. Morse invented the telegraph, which allowed messages to be transmitted over wires. The concept of his invention was that when electricity flows in a wire, it can be detected and converted to sound; when there is no flow of electricity, there is no sound. It was a simple on/off system. However, there had to be a method of making sense out of the sound/no sound feature.

Morse devised a code consisting of dots and dashes, which today is known as the Morse code. If the transmitter is turned on for an instant, the result is a dot. If it is turned on for a longer time, the result is a dash. Each letter of the alphabet has a code, such as "dot, dash" for the letter *A*, and "dot, dot, dot" for the letter *S.* When transmitting the code, each letter is separated by a time interval equal to three dots, and each word is separated by a time interval equal to seven dots.

During its day, Morse's system was praised as "the instantaneous highway of thought."

Some time later, Guglielmo Marconi invented wireless

telegraphy, a precursor to our modern radio communication. Because it transmitted a single tone, it required far less power than voice transmission and so could be sent over much greater distances. The signal was simply turned off or on to follow the code invented by Morse.

Although most authorities credit Marconi as the inventor of the radio, in 1943 the Supreme Court of the United States ruled that Marconi's patents were invalid due to Nikola Tesla's previous descriptions.

In the early 1900s many wireless telegraphy operators on ships were former railroad or postal telegraphers. If an operator wanted to send out one message and make sure that all stations heard it, he would begin the message with the letters *CQ*, which meant "all stations." The operator would do this when sending out time signals or other general notices.

In 1904 it was suggested that *CQD* should be used as a distress signal. In other words, "all stations, distress." A few years later, the Berlin Radiotelegraphic Conference brought up the subject of an international distress signal. After a lengthy discussion, it was agreed that SOS would be the new distress signal. Participants thought that if three dots, three dashes, and three dots were sent as a single string, it could not be misunderstood.

FACTOIDS

The first recorded use of the SOS distress signal by an American ship was in August 1909, when the SS *Arapahoe* radioed for help after losing its screw off Cape Hatteras, North Carolina. A few months later, the *Arapahoe* picked up an SOS signal from the SS *Iroquois*. The *Arapahoe*'s radio operator was thus the first person to both send out an SOS from an American ship and to receive an SOS from an American ship.

Even after the invention of the telephone in 1864, the telegraph was the world's primary form of telecommunication for over 50 years.

When the Titanic sent out its first distress signal, it was *CQD* followed by *MGY,* which were the *Titanic*'s call letters. After sending out *CQD* a number of times, the radio operator then sent an *SOS.* Subsequent calls were *CQD*'s interspersed with *SOS*'s.

Morse code could be considered the precursor of modern computer codes. The telegraph signal was either on or off. A modern computer works in the same way, using a binary code consisting of a "1" (on) or "0" (off).

The inventor of the Morse code, Samuel Morse, was not an engineer. He was a Massachusetts portrait painter.

If you ever visit the U.S. Capitol building, be sure to look at the Rotunda. One of the figures in the center of Italian artist Constantino Brumidi's beautiful fresco is none other than Samuel F. B. Morse.

DID YOU KNOW?

The code that Morse created in 1832 died a quiet death 165 years later. In 1997 Morse code ceased to be the official international language of distress, being replaced by much more sophisticated satellite-based "Mayday" electronic systems. (Mayday is derived from the French *m'aidez,* which means "help me.")

Morse code may be dead, but it's not buried yet. Amateur radio operators use it quite often, especially in times of disaster when other forms of communication are not available.

The military also maintains a Morse code capability. Billion-dollar satellites can malfunction or be jammed and sophisticated ground networks can break down during a battle. As a contingency, every year the U.S. army trains 2,800 soldiers to become proficient in Morse code. Every U.S. merchant ship must have on board a radio officer who can transmit and receive Morse code. In fact, while at sea the officer must spend eight hours a day monitoring the radio for Morse code distress calls.

Many military messages end with the phrase, "Over and out."

Well, Morse code may be over, but it's not out.

What volcanic eruption killed the most people in the twentieth century? (Some politicians will do anything to get a vote.)

On May 8, 1902, Mount Pelée on the island of Martinique completely destroyed the city of Saint Pierre and killed 29,000 people. Martinique is a volcanic island in the Caribbean Sea, about 400 miles northeast of Venezuela.

Mount Pelée started rumbling in early April and spewed out clouds of gray ash until the narrow streets of Saint Pierre were covered with layers of it. Governor Mouttet persuaded the local newspaper to downplay the danger. An election was coming and the governor wanted to make sure that the white and wealthy leaders stayed in power. The wealthy would be the first to leave the island if there was danger, and with them gone, the ruling party could lose the election to the black inhabitants of the island.

A minor eruption blew a crack in the volcano on May 3, and ash and mud destroyed a mountain village. Although the American consul sent a telegram to Washington, D.C., the governor intercepted it and sent a different telegram saying that the danger was over.

The danger was far from over. The weight of the raining ash collapsed roofs all over the city. Hundreds of people who lived close to the volcano were killed and those who survived sought refuge in Saint Pierre. The population of the capital grew to about 30,000 people.

The governor finally realized that the danger from the volcano was real and planned to evacuate the city the following day, but he was too late.

Mount Pelée erupted at 7:59 A.M. the morning of May 8. A searing black cloud of volcanic gas, with temperatures hotter than 1,300°F, raced down the mountain slopes, picking up debris as it went, at speeds exceeding 60 mph.

Three minutes later the Saint Pierre telegraph operator sent his last message. It took only seconds for Saint Pierre to burst into flames. The intense heat melted glass and steel. The hot volcanic ash suffocated the people, and all 30,000 residents, including Governor Mouttet, were killed.

Just one minute later a ship at sea sent the message "Saint Pierre destroyed by Pelée eruption. Send all assistance."

Rescue parties found only three survivors. One woman had taken shelter in a ditch outside of town and a cobbler had sought refuge in his basement. Both were burned and bleeding, but they survived.

It took search parties three days to find the third survivor, a black man who had been sentenced to death for the murder of a white Frenchman. He was to be hanged the day the volcano erupted. Rescuers found him in an underground jail cell.

It is ironic that the 30,000 people who wanted him hanged all died, while the accused man was the only one in the city to live. He was later given a pardon.

FACTOIDS

The eruption of Mount Pelée was the third deadliest volcanic eruption in the past thousand years.

When the Indonesian volcano Tambora erupted in 1815,

92,000 people died. However, it wasn't the volcano that killed them. The ash destroyed the land and the people died of starvation.

Approximately seven thousand years ago, a volcano named Mazama erupted in what is now Oregon, and the falling ash covered the entire northwestern United States. The explosion was so violent that the mountain collapsed and left a crater six miles in diameter and half a mile deep. It eventually filled with rain water and is known today as Crater Lake, Oregon.

DID YOU KNOW?

In Mexico there used to be a village called Paricutin. It was near a village called San Juan Parangaricutiro. Neither village exists today. What destroyed them started in a nearby cornfield on February 19, 1943.

On that day, Dionisio Pulido was working in the cornfield, but came back to the village claiming that the field was burning his feet. Fortunately, the village priest realized what was happening and had the people evacuate the area.

By the next morning, the once-flat cornfield was a small hill, 100 feet high. The hill was erupting violently, sending fire and ash into the air. In the first year the hill, now an active volcano, grew to 1,100 feet. Although it continued to grow for another eight years, it only added 245 feet and today is 1,345 feet high.

The slow-moving lava from the volcano eventually covered a 10-square-mile area. The villages of Paricutin and San Juan Parangaricutiro were buried in lava except for the church steeple, which can still be seen today.

Only three people were killed, none by lava or ash. They were all killed by lightning generated by the volcano. It is quite common for lightning to form in the ash column from an erupting volcano. The friction among ash particles causes an electrical charge to form within the eruptive cloud. This electrical charge builds up until it eventually discharges, producing a lightning bolt.

The eruptions stopped in 1952. Scientists from around the world had come to watch the birth of a new volcano, something no one had ever seen before. Although they were ecstatic, it was no consolation to Dionisio Pulido. A towering volcano now stood where his lovely, flat cornfield had once been.

What is the largest earthquake ever recorded in the continental United States? (Don't get shook up when you read this.)

If you guessed that the largest recorded earthquake in the history of the United States was in California, that's a very good guess. Unfortunately, it's wrong. The correct answer is Missouri.

The *New Orleans,* the first steamboat on the Mississippi River, was on its maiden voyage. On the evening of December 16, 1811, the captain tied his ship up to an island in the river and went ashore for the night with his crew. When they were abruptly awakened a few hours later, the steamboat was gone, and so was the island. They had been submerged by the most powerful earthquake ever to hit the continental United States. It was near the town of New Madrid, Missouri, and became known as the New Madrid earthquake.

Actually, three earthquakes occurred in late 1811 and early 1812. The only part of the country that didn't feel them was the Pacific Coast. The first quake was so powerful that it rang church bells in Boston, some 1,000 miles away, damaged buildings in Washington, D.C., knocked over chimneys in Maine, and permanently changed the course of the Mississippi River, which flowed backward at one point.

Because of the earthquakes, new lakes were formed and forests were devastated over an area of 150,000 acres. Most houses in New Madrid were destroyed and many of them, including gardens and fields, were simply swallowed up. Survivors said they had seen the ground rolling in waves, huge cracks opening up in the earth, and large areas of land either sinking or rising. There were also reports of tornado-like winds and geysers of mud and rock.

In spite of the severity of the earthquakes, Missouri was only sparsely settled in the early 1800s, and damage and fatalities were minimal.

Not many people know about the New Madrid earthquakes. However, scientists have studied them and know that they occurred in a seismic zone that covers an area of roughly 1,200 square miles. Today millions of people live on top of this earthquake zone, many in large cities such as Saint Louis, Missouri, and Memphis, Tennessee. If there is ever another New Madrid earthquake, damage and death will certainly not be minimal.

If you're worried about earthquakes, there's one place you can move that has the least number of earthquakes of any place on earth: Antarctica. If you decide to move there, be sure to dress warmly.

FACTOIDS

Both the Greek philosopher Aristotle and William Shakespeare believed that violent winds were trapped in huge underground caverns. Earthquakes were caused by the struggle of the winds trying to escape from the caverns.

Today the central Mississippi valley has more earthquakes than any part of the United States east of the Rocky Mountains.

The Richter scale used to measure earthquakes is not like a typical measuring device. For example, an earthquake of magnitude of 9.7 is about 23,000 times stronger than an earthquake of magnitude 6.8. The first New Madrid earthquake had a magnitude of 7.8.

The New Madrid earthquakes are among the great earthquakes in known history. No other earthquake on the North American continent has changed the topography of the land as much as they did.

The largest earthquake in the world, with a magnitude of 9.5, since 1900 occurred in Chile in 1960. It killed 2,000 people, injured 3,000, and left 2 million homeless. It also caused damage in Hawaii, Japan, the Philippines, and along the west coast of the United States.

DID YOU KNOW?

A man predicted the New Madrid earthquakes. Ten years before the earthquakes occurred, he said, "In the midst of the night the earth will begin to tremble, giant trees will fall, rivers will run backward, new lakes will be formed, and old ones will disappear." On the exact day that he predicted, the New Madrid earthquakes shook the eastern half of the United States.

The man was Tecumseh, a Shawnee warrior and chief. His father, Pucksinwah, saw a meteor when his son was born and named him "Panther Crossing the Sky," or Tecumseh.

Tecumseh realized that all Native Americans needed to unite as one, and that their differences were petty compared to the problem of the white settlers who threatened the Native American way of life. He united the tribes of the Ohio Valley and the Great Lakes to soundly defeat well-armed armies of the United States.

To Tecumseh, the New Madrid earthquakes were a signal for all tribes to unite against the whites. Unfortunately, Tecumseh left the area to enlist more warriors and put his brother in charge. This was a mistake. The Native Americans were defeated in the battle of Tippecanoe.

Tecumseh's confederation of tribes collapsed, and he joined the British army during the War of 1812. He was killed in the battle of the Thames. The man who fired the bullet was Colonel Richard Johnson, who later became vice-president of the United States.

If some dark summer night you see a flash of light across the sky, watch it carefully. It just might be the "Panther of the Sky."

What was the deadliest fire in U.S. history? (It wasn't Chicago or San Francisco.)

On October 8, 1871, a forest fire swept across the town of Peshtigo, Wisconsin, destroying 1.2 million acres of timber and killing an estimated 1,200 people. Although it was the worst fire in the history of the country, it was quickly overshadowed by the great Chicago fire, which took place on the same day and year. Although only 250 people were killed in the Chicago fire, it destroyed over 17,000 buildings and received a great deal of publicity, while the Peshtigo fire did not.

Peshtigo was a boom town that sprung up near a thriving lumber mill. The spring and summer of 1871was unseasonably dry and hot, and many creeks dried up. By mid-September, sporadic fires broke out near the town and peat fires smoldered for three weeks or so. October 8 was a Sunday, and the residents of Peshtigo were not overly concerned about the dense smoke hovering over them. That evening, a sudden wind whipped up the smoldering peat, and by late evening the wind was as strong as a tornado. Sheets of fire leapt 100 feet into the sky.

In just minutes the fire was rolling over the ground and the air itself seemed to be on fire. The fire consumed all of the oxygen so there was none left to breathe. People tried to flee but couldn't outrun the racing fire. In just 20 minutes, all that could be seen for miles around was billowing smoke and dancing flames. The heat was so intense that it split huge boulders.

A few people managed to reach the Peshtigo River, but there weren't many survivors. The entire town was burned to the ground in less than an hour. All that remained were the charred bodies of 800 Peshtigo residents. Combined with the deaths from the fire in other small towns, the total number of fatalities was 1,200.

Like a phoenix rising from the ashes, Peshtigo was eventually rebuilt. Perhaps in compensation for its terrible tragedy, it was one of the very few cities in the United States that was untouched by the Great Depression.

FACTOIDS

Although many people survived the fire by staying in the river, all the fish in the same river died.

The town's Catholic church was just a charred ruin, but the tabernacle containing the sacred vessels was untouched and later found on the river bank.

John Mulligan, although burned in the fire, walked six miles to the neighboring town of Marinette and got help for the victims of the fire.

A man who was caught looting corpses was found guilty by a hastily formed jury of survivors and sentenced to be hanged. However, no one could find a rope to hang him with because all the rope had burned up in the fire, so they decided to set him free.

Only one building in Peshtigo survived. It had been built with green lumber which contained enough moisture to keep the building from igniting. The building is still standing in Peshtigo today.

DID YOU KNOW?

There have been many forest fires in the history of the United States. One of the largest and most destructive fires of the twentieth century, in Tillamook, Oregon, started the "six-year jinx."

The fire started on August 14, 1933. It's believed that a spark from a steam-powered engine used to haul logs started the blaze. As the fire grew, a column of smoke 40 miles wide mushroomed 40,000 feet into the air and could be seen throughout the Northwest and for hundreds of miles out at sea. At its peak, there was a 15-mile front of flames, and the fire burned 10,000 acres an hour for over 20 hours. The fire was not completely out until 19 days after it had started. It had consumed 239,000 acres, an area about half the size of Rhode Island.

Fires struck the same area three more times. The first time was on August 1, 1939, almost exactly six years from the date of the original fire. It destroyed 189,000 acres, 19 million feet of cut logs, and logging camps and equipment.

The jinx struck again, six years later, in July 1945. It took 3,500 men six days to control the fire, which burned 180,000 acres.

Six years later, on July 20, 1951, the jinx struck again. This time the fire was controlled in five days and only burned 37,000 acres that had been burned in the previous fire.

All four fires consumed over 13 million board feet of timber, enough to build over 1 million five-room houses.

Fortunately, the six-year jinx ended in 1951. If it should return, it won't be until 2005. Hopefully, by then everyone will be more careful when camping, smoking, or logging in the Tillamook forest.

How many aircraft crashes have occurred in scheduled airline flights in the past fifty years? (Next time try the train.)

Airlines in the United States have an enviable safety record. Fatal accidents per million aircraft miles flown are usually less than one in a thousand. In other words, there is only one fatal accident for every billion miles flown. However, these accidents make the news because they are spectacular and result in so many fatalities at once.

In the past 50 years there have been 258 aircraft accidents, or an average of a little over 5 per year. Some years have been worse than others. There were 10 major accidents in 1959, 12 in 1960, and 13 in 1968. On the other hand, there were none at all in 1980 and only one each in 1984 and 1993.

The worst death toll in aviation history occurred in the Canary Islands in 1977. A Pan Am plane was on the runway in a heavy fog when a KLM plane started its takeoff without permission. The KLM plane collided with the Pan Am plane and both burst into flames. The 560 passengers and 23 crew members on both planes were killed. There were no survivors.

The worst disaster involving only one plane occurred in 1996 when a Japan Air Lines Boeing 747 experienced control problems and crashed into a mountain in Japan. There were 520 fatalities.

The third worst disaster, claiming 349 lives, also occurred in 1996, the midair collision of a Saudi Arabian Airways Boeing 747 with a Kazakhstan Airlines Ilyushin I1-76TD. Indications pointed to pilot error in the Kazakhstan Airlines plane.

Each of these disasters represents one of the three major factors in airline crashes: weather, mechanical failure, and pilot error.

Not all aircraft crashes result in death. In 1997 a United

Airlines Boeing 747 crashed into the Pacific Ocean. Only one of the 374 people aboard died.

FACTOIDS

In addition to weather and mechanical failure, many airline crashes have been caused by bizarre events. Here are just a few.

A Brazilian aircraft crew was so preoccupied with listening to the World Cup soccer match that they flew in the wrong direction, then ran out of fuel and crashed in the jungle. The survivors walked two days through the jungle to reach safety. It's alleged that the pilot's very first words when he emerged from the jungle were "Who won?"

A Pacific Southwest Airlines plane went into a steep dive and crashed after an employee who had recently been fired shot and killed both pilots.

The captain of a Russian International Airlines plane let his children (11-year-old daughter and 16-year-old son) take turns flying the plane. The boy put the aircraft into a steep bank and the plane stalled, went into a spiral, and crashed. All aboard were killed.

The pilot of a British Overseas Airways flight wanted to give his passengers a good view of Japan's Mount Fuji. The plane encountered severe clear air turbulence and crashed into the mountain. Everyone aboard was killed.

When a bomb exploded in the forward cargo hold of a JAT Yugoslav Airlines plane, the tail section broke off with a flight attendant still in it and fell 33,000 feet. Although the flight attendant broke her legs and was paralyzed from the waist down, she survived.

DID YOU KNOW?

Talk of aircraft disasters always brings to mind the talented people whose lives have been cut short. Some have died during commercial flights, others during private plane accidents.

Entire sports teams have been killed in commercial airline disasters, including the U.S. Olympic boxing team, the U.S. figure skating team, England's Manchester United soccer team, and the Evansville, Illinois, basketball teams. The football teams of Marshall University, Wichita State College, and Cal Poly at San Luis Obispo, California, were also victims of commercial airplane disasters.

Perhaps because entertainers and sports personalities travel so much, they often use private planes or chartered aircraft. We've lost many wonderful singers in airplane crashes. Some of these include John Denver, Rick Nelson, Otis Reading, Jim Reeves, Ritchie Valens, Patsy Cline, Buddy Holly, and Jim Croce. In Don McLean's song "American Pie," he is referring to the plane crash in 1959 that killed Buddy Holly, Ritchie Valens, and J. P. Richardson ("The Big Bopper") when he sings about "the day the music died."

We've also lost sports heroes, including baseball players Thurman Munson and Roberto Clemente, and boxing champions Rocky Marciano and Marcel Cerdan.

Governor George Mickelson of South Dakota, Senator John Tower, U.S. Commerce Secretary Ron Brown, and U.N. Secretary Dag Hammerskjöld all died in airplane accidents.

Airlines carry people from all walks of life and a disaster can affect them all.

More questions? Try these Web sites.

Historic Disasters

http://members.xoom.com/generanch/disaster/disaster.html

If you're interested in disasters of all types, this site is for you. It covers natural disasters such as earthquakes, blizzards, droughts, epidemics, hurricanes, and famine as well as famous air disasters, bridge collapses, and railroad disasters.

Wrecks

http://members.aol.com/roxIn/ships/wrecks.html

This site describes nine famous maritime disasters including the *Andrea Doria* and the *Titanic*. It includes links to other sites having information about the specific shipwreck.

Vessel Casualties and Pirates Database

http://www.cargolaw.com/presentations_casualties.html

Although this is a commercial site, it has a great deal of information about maritime disasters and piracy. Scroll down the page to see links such as "U.S. Shipwreck Index 1828–1911" and "Modern Day Pirates."

Volcano World

http://volcano.und.nodak.edu/

There is probably more information about volcanoes on this site than any other. It is updated constantly and lets you know about all current eruptions. You can even add your name to a mailing list to receive e-mail notices of all new volcanic activity.

Click on "Volcano adventures!" for a list of various expeditions to volcanoes. Click on any expedition to read the full story and see the wonderful photographs.

Because this site has so much information, you should take some time to explore all of its features. It's well worth the effort.

The Top 10 Killer Tornadoes

http://tornadoproject.com/toptens/topten3.htm

This site has a table of the top 10 killer tornadoes in U.S. history. If you scroll down, each tornado is described in detail.

The very bottom of the page has links to many tornado Web sites, including tornado myths, tornado oddities, frequently asked questions (FAQ) about tornadoes, and tornado safety.

Earthquakes

http://wwwneic.cr.usgs.gov/

An excellent earthquake site. It includes information about current earthquakes, large earthquakes of the previous year, and links to other sites with earthquake information.

5

Far Out

What was the world's largest typewriter? (Do giants know how to type?)

For years, the world's largest typewriter attracted huge crowds at the Garden Pier in Atlantic City, New Jersey.

The gigantic typewriter was built by the Underwood Company to promote its line of typewriters. It was so large that the company often had a young lady sitting on each one of its keys.

Although it was an extremely popular attraction as well as a wonderful advertisement for the Underwood Company, the huge machine was destroyed because of a war.

Metal was in short supply during World War II, so the

company scrapped its giant typewriter to help the war effort. We probably would have won the war even if they had not melted down the typewriter.

The Underwood Company had a long history of manufacturing typewriters. In 1899, the company produced the first typewriter that allowed keys to hit the paper and then fall back into place so the typist could see the words as they were typed. All previous machines used a "down" strike method in which the keys hit the back of the paper. This was called "blind typing" because the typist could not see the words.

At one time, speed typing contests were all the rage and were treated as a sport. An Underwood employee, Charles E. Smith, trained the Underwood typists like athletes, often working them eight hours a day, five days a week. In spite of stringent rules covering line length and the consistent darkness of all letters, some of his students could type 140 words a minute on the cumbersome manual typewriters. Smith was so successful that he actually killed the popular typing contests. Underwood's typists had an unbroken string of victories. Because they always won every contest, the public lost interest in the competition. In this case, success caused the failure of a great publicity stunt.

FACTOIDS

Around 1836, one enterprising inventor built a typewriter as large and ungainly as a pinball machine.

The typewriter ampersand symbol (&) was invented by Marcus Tullius Tiro around 60 B.C. as part of a system of shorthand that allowed him to write down the orations of Cicero. The symbol was a combination of the letters *e* and *t* from the Latin word *et* meaning "and."

The fastest typing speed on a manual typewriter, 176 words per minute, was achieved by Carole Bechen in 1959. In 1949,

Stella Garnand typed 216 words a minute on an IBM electric typewriter.

The familiar typewriter @ symbol is simply called the "at" symbol in the United States. However other countries have given it more descriptive names such as monkey's tail, elephant's ear, and cinnamon bun.

One of the first manufacturers of typewriters was Remington and Sons, which previously manufactured guns and sewing machines. The first typewriter they produced had a foot pedal, similar to that on a sewing machine, to advance the paper.

DID YOU KNOW?

Although the first typewriter was patented in 1714 in England, the inventor never bothered to actually build a machine and the details of his design have been lost forever. It was almost 100 years later when Pellegrino Turri invented and built typewriters in 1808. None of his machines has survived, but some documents created by them still exist.

Prior to 1860 none of the typewriters had keyboards. They had a selector dial for choosing a letter and a lever that was pulled to make the impression on the paper. An American, Christopher Latham Sholes, invented a typewriter with a keyboard.

Typists using the newly invented keyboard used either two or four fingers, looked at the keyboard, and used the "hunt-and-peck" system to find the letter they wanted to type. Some people prefer to call this "the biblical system," or "seek and ye shall find." Mrs. M. V. Longley had a better idea and developed an "all-finger" method. This idea was revolutionary at its time and led to the invention of the "touch-typing" system by Frank McGurrin. In 1888 McGurrin competed with Louis Taub, the champion four-finger typist. McGurrin easily won, and his new touch-typing system soon spread throughout the country and the world.

Typewriters have come a long way since then. The IBM Selectric used a movable-type ball instead of levered keys, and modern computers handle everything electronically.

Yet with all of the advances, we are still using the same keyboard design that was invented in the 1860s. In spite of many attempts to change the keyboard layout, it is still the same.

Typewriters may soon become obsolete. But no matter what type of machine or computer we use in the future, there's a good chance that the top keyboard line will still be QWERTYUIOP!

Who was the model for the Gerber baby? (No beanie on this baby.)

For over 70 years an angelic baby face has been the picture adorning jars of Gerber's baby food. For many of those years people have guessed that the Gerber baby grew up to be anyone from Elizabeth Taylor to Bob Dole. In fact, there was even a lawsuit that challenged the true identity of the baby.

The most common guess is that the symbol is a picture of Humphrey Bogart when he was a baby. Bogart's mother was a commercial illustrator and drew a picture of Humphrey Bogart as a baby which was used in advertisements for a baby food company. However, that happened a long time before the Gerber baby. In fact, Humphrey Bogart was 29 when Gerber introduced their baby picture.

In 1928 Gerber was preparing an ad campaign to introduce its new baby food and decided to use a baby's face. Leading artists of the time submitted their works, most of which were elaborate oil paintings. One artist, Dorothy Hope Smith, submitted a charcoal sketch of a baby and asked if it was about the right age. She offered to finish the drawing if the company liked it.

The simple drawing so impressed the executives that they

decided to use it just as it was. It became so popular that three years later the company used it as its trademark. It's still being used today.

For many years the identity of the model was kept secret. Finally, the model herself granted newspaper interviews. Dorothy Hope Smith's drawing was a sketch of mystery novelist and retired English teacher Ann Turner Cook, who was only five months old at the time the drawing was made. She was two years old when the drawing was selected as the Gerber baby.

The original charcoal sketch still exists. It is safely locked in the company's vault.

FACTOIDS

Many polls have been taken to select the most important babies of the 20th century. The Gerber baby was on virtually every one. Here are some others.

Charles Lindbergh, Jr. Son of the famous aviator, he was kidnapped and later murdered. His death made the public aware of child safety concerns.

Baby Fae. She was born with an incurable heart disease. Doctors replaced her heart with the heart of a baboon. Although she later died, her surgery helped advance the treatment of infant heart disease.

Little Ricky Ricardo. The *I Love Lucy* show was the first program to portray the trials and tribulations of pregnancy, birth, and raising a small child.

Billionth Chinese baby. The Chinese government used this birth to institute stringent birth control laws to stem population growth.

James Hathaway. His father watched his birth in the hospital, something unheard of at the time. Eventually fathers were allowed in the delivery room, a common practice today.

DID YOU KNOW?

Perhaps the most famous babies of the 20th century were the Dionne quintuplets. Although multiple births are common today, when the Dionne children were born in 1934, they were the only living quintuplets in the entire world. The five girls were named Annette, Emilie, Yvonne, Cecile, and Marie.

Worldwide interest was so great that the Canadian Parliament took them from their parents and made them wards of the king. Although the reason was ostensibly to safeguard the children's health and protect them from exploitation, the government did just the opposite. The quintuplets were moved into a special facility called "Quintland," which was designed as a theme park to attract tourists. The children were displayed as a curiosity to adoring crowds of people three times a day. Always in the public eye, they never had a normal childhood. The scars of abuse and exploitation stayed with them throughout their lives.

Although the five girls accounted for the largest amount of tourist dollars in Canada, estimated at around $500 million, neither they nor their parents ever saw a penny. They were not returned to their parents until they were nine years old.

In recent years there have been births of sextuplets, septuplets, and even octuplets. Fortunately, the parents kept all of their children.

Medicine has come a long way since 1934, and so has the understanding of what's proper for a child's welfare. Society has learned that children should not be displayed as curiosities, nor should they be abused and exploited.

Who was Prester John? (Was Marco Polo Looking for him?)

In 1145, Western Europe was preparing for the Second Crusade. The Turks had become a serious threat to Christian kingdoms in the Holy Land, and the Crusaders were seeking help. Then it happened. A rumor swept through Europe about a priest-king named Prester John. (Prester is a corrupt form of *presbyter,* which means "priest.") It was said that he had defeated the Medes and Persians and had continued his advance to link up with the Crusaders. Unfortunately, he had to turn back when he reached the Tigris River because he had no boats.

This mysterious king was supposedly descended from the Magi, the wise men who visited the Christ child. Even more appealing to the Crusaders was the story that his army consisted of almost 1.5 million soldiers and over 125,000 horses.

A letter, reported to be from Prester John himself, was delivered to the Byzantine emperor and circulated to Pope Alexander III and Frederick Barbarossa, the Holy Roman emperor. The writer called himself the King of Kings and described the wonders of his kingdom, which included white and red lions, white bears, centaurs, giants, the ruins of the Tower of Babel, mysterious pebbles that could cure the blind or make a person invisible, and even the fountain of youth.

Although the Byzantine emperor decided the letter was pure fantasy, many Western Europeans believed it was factual. Pope Alexander III himself sent a personal emissary to the East bearing a letter for Prester John. The emissary was never heard from again. Still, the legend persisted for over 500 years.

It's easy to see why people wanted to find Prester John. According to his letter and other rumors, his kingdom featured Amazons, one-eyed giants, gold, silver, and precious stones. He was served by 72 kings, all of whom were bishops or abbots. There was no vice, crime, or poverty anywhere in the kingdom.

For 500 years Europeans sought to find the kingdom of Prester John. In the process, they explored the Far East, Central Asia, and Africa, establishing contacts in countries they didn't know existed and expanding their knowledge of the world and its peoples.

Nevertheless, no one ever found the kingdom of Prester John.

FACTOIDS

The legend of Prester John first appeared around 1145 during the Second Crusade, but his kingdom was still included on maps of the world as late as the 1600s.

As time wore on, new editions of Prester John's letter became more fanciful. One of them described a salamander that could live in fire. In actuality, it was asbestos.

The first recorded mention of a fountain of youth was in the letter from Prester John. It was mentioned again 93 times in various documents written about Prester John.

Today, libraries in Europe have over 100 manuscripts telling the story of Prester John. They are written in many different languages, including Hebrew. The original letter was written in Latin.

DID YOU KNOW?

When people think of the fountain of youth, they usually don't think of Prester John, but of Ponce de León, a Spanish explorer who searched for it.

Ponce de León accompanied Columbus on his second voyage to the Americas and settled on Hispaniola, an island in the Caribbean. There he was governor and military commander for nine years. He also discovered another island close by. It was named Boriquén by the natives, but today we know it as Puerto Rico. Although he had wealth, power, and glory, he wanted more. After the king gave him more men and ships, Ponce de León set out on his quest for the fountain of youth. In his search he discovered a land that he named "Pascua de Florida" or "the feast of flowers" after a Spanish Easter holiday. This is now the state of Florida. He died without ever finding the fountain of youth.

Historians all agree that the fountain of youth is nothing more than a myth. But have people truly given up searching for it? Not at all.

Tibetan monks are said to have a series of exercises to prevent aging. There are claims that a British naval officer who did the exercises took 30 to 40 years off his body, as is described in Peter Kelder's book *Ancient Secret of the Fountain of Youth.*

In India, ancient Ayurvedic techniques supposedly helped a mahatma live to be 185 years old, and some people today claim that these techniques can extend anyone's life.

Today many countries have numerous cosmetic products to prevent or disguise the effects of aging.

In 1998 scientists succeeded in producing cells that live twice as long as normal cells and, because they slow down the aging clock, might someday be used to allow people to live much longer and healthier lives.

It seems that even today people are still searching for the fountain of youth.

Who was the first Ronald McDonald? (Don't be a bozo when you answer this.)

The first Ronald McDonald appeared on television in 1963 and was portrayed by Willard Scott. There's a good chance you've seen Willard Scott without his Ronald McDonald makeup. He began reporting the weather on the *Today* show in 1980. Every day he wishes happy birthday to people over 100 years old. Since 1987 he has anchored the annual Macy's Thanksgiving Day parade. He has also lit the national Christmas tree in Washington, D.C., for over 30 years, and has been the White House Santa Claus.

The popular Scott has appeared in Carnegie Hall, where he narrated "The Night Before Christmas." He has also appeared at the Palace Theatre and the Grand Ole Opry.

In the four years before being hired to portray Ronald McDonald, Willard Scott was Bozo the Clown in Washington, D.C.'s *Bozo's Circus* television program.

Although Willard Scott was the first Ronald McDonald, the one you are most likely to remember is King Moody, who played Ronald McDonald on television for 18 years. When some 107 Ronalds met at the first worldwide Ronald McDonald gathering, they wanted to know how to answer the question children often asked, "Are you the real Ronald McDonald?"

In a stirring talk, King Moody explained that he was only an actor who played the part. Without the children, Ronald was simply a piece of film; he supplied Ronald's face but they, the children, supplied Ronald's heart.

King Moody became a king among his peers that day.

FACTOIDS

When it comes to recognition, Ronald McDonald is second only to Santa Claus.

When he was 52 years old, Ray Kroc mortgaged his home and spent his entire life savings to become the exclusive distributor of a five-spindled milkshake mixer. He had heard of a place in California that was using eight of these mixers at a time, so he headed west to see this hamburger stand. When he arrived he was amazed to see so many people served so quickly. He convinced the owners, Dick and Mac McDonald, to open up more restaurants. His plan was to sell eight of his mixers to each of the new restaurants. At the time he didn't realize he had started what was to become one of the largest restaurant chains in the world.

The oldest McDonald's is in Downey, California. It opened

in 1953, the year before Ray Kroc was given exclusive franchising rights by the McDonald brothers. A year later, he opened his first store in Des Plaines, Illinois. Today this store is now a museum, which is a replica of the first McDonald's restaurant opened by the founder of McDonald's.

A professional football player with the Philadelphia Eagles, Fred Hill, was responsible for the first Ronald McDonald house. In 1974, his daughter was being treated for leukemia. He and his wife grew tired of trying to sleep on hospital chairs and benches and eating out of vending machines. They saw other parents who could not afford the cost of staying in hotels suffering the same fate. He enlisted the aid of his teammates, the club manager, a doctor from the Children's Hospital of Philadelphia, and the local McDonald's restaurants. Together, they created the first Ronald McDonald house. Today, there are now almost 200 such houses in 16 countries.

DID YOU KNOW?

King Moody did more than play Ronald McDonald on television. He was an accomplished actor who appeared in many films, numerous television shows, and over 1,000 television commercials.

One such show was *Get Smart* which ran from 1965 to 1970. It was a comedy spoof of spy shows whose hero, Maxwell Smart, was a bumbling spy who worked for an agency so top secret that even the CIA didn't know it existed.

Smart worked for CONTROL (the good guys) and was always fighting the evil KAOS (the bad guys). These names are not acronyms. The writers felt that evil represented chaos, so they named the evil team KAOS. The opposite of chaos is control, so that's the name they gave to the good team.

King Moody played the henchman to the main KAOS

agent, Ziegfried. He also played a bad guy on *Police Woman* with Angie Dickinson.

King Moody has not done any acting since 1992. The last news about him was that he was writing a book about his experiences playing Ronald McDonald.

Hopefully, King Moody will finish his book. It would be a wonderful story to share with all of his Ronald McDonald fans.

Have scientists actually been able to teleport an object from one location to another location? (Beam me up, Scotty.)

In the *Star Trek* films and television series, a person in one location was dematerialized, sent elsewhere on some type of electronic beam, and rematerialized in another location. In effect, the person was "teleported," or moved, from one location to another in just a few seconds. Teleportation is the process of transporting an object from one place to another without touching it or using any mechanical device. In true teleportation, the object disappears from where it is and later reappears in some other spot. Is this possible, or is it just science fiction?

In 1997 scientists in Austria managed to achieve what they called teleportation. Later, scientists in Rome and California also achieved teleportation. The California researchers claimed they had achieved the first true teleportation. Instead of a physical object, they moved light. In a tabletop experiment, the scientists encoded a light beam and used it to generate an electrical current. The current was then sent to the other end of the table where it was decoded to produce a replica of the original light beam. They did not physically transport the beam itself but sent its properties to another beam, thereby creating the replica.

Although scientists believe that teleportation techniques can be used in the next few decades for transmitting information, they think that the ability to teleport people is virtually impossible in our lifetime because of the vast amount of information that would have to be assembled and transmitted.

If it were possible to copy the atoms in a human being, the amount of information that would have to be transmitted from one place to another is mind boggling. With the best fiber optics we have today, it would take over 100 million *centuries* to transmit the information.

Perhaps walking isn't the worst way to get from one place to another. In fact, it just might be faster.

FACTOIDS

The term teleportation was coined in 1931 in Charles Fort's book *Lo!*

Some scientists believe that within 10 years they will be able to transport a virus. But why a virus? Do people need to have the flu teleported from one place to another?

Teleportation techniques are already being applied to quantum cryptography, which is a virtually impenetrable method of communicating information. If anyone attempts to intercept an encrypted message, the message will immediately self-destruct. There is already a quantum computer in a laboratory in Los Alamos, New Mexico.

When experimenting with teleportation, whether it is a beam of light or a particle, for some unknown reason the researchers always call the sender "Alice" and the recipient "Bob."

DID YOU KNOW?

Another way of moving objects from one place to another without touching them or using any mechanical device is called psychokinesis, or mind over matter.

Instances of mind over matter have been recorded since ancient times. Usually the phenomenon has been attributed to mystics, holy men, magicians, and even devils. Even in more modern times, the idea of mind over matter has been associated with the occult or with fraud. For that reason, few scientists took it seriously.

In recent years, scientists have been taking psychokinesis very seriously. In 1991 researchers at Princeton University used elaborate electronic devices to determine if subjects could control the rolling of dice with their minds. The laboratory used a device that provided an electronic roll of the dice.

The results of the study were positive. They showed that a person could mentally control the roll of the dice, and distance did not matter. Even if the subject were 1,000 miles away, the dice could still be controlled by the mind.

Today, scientists think they can no longer ignore psychokinesis as fraudulent but must study it if they are to learn more about it.

Teleporting an object to some distant location is not yet possible. Controlling the throw of the dice with the mind can still be accomplished by only a select few.

It may not be teleportation, but if you really must move an object from one location to another, you can always pick it up, carry it to the new location, and set it down. It's guaranteed to work.

Is it true that fleas have been trained to perform circus acts? (You're probably itching to hear the answer.)

People have been training fleas for hundreds of years and have used them to perform various acts in what is known as a "flea circus." The flea circus originated in England in the 1600s. Flea circuses reached their height of popularity in the 1830s largely because of the efforts of Signo Bertolotto, who toured Europe with his performing fleas. His fleas danced, pulled coaches, and even wore costumes. However, the fleas were not at all happy about being in a circus. They were permanently glued to each other, or to the props, or to the miniature chains that let them pull wagons.

Around 100 years ago, flea circuses were a favorite parlor entertainment throughout Europe and were an attraction at numerous county fairs in the United States.

There are still many flea circuses today. However, fleas are no longer cruelly glued to props. One method of training them is to use tuning forks with different frequencies. The sound of one fork might be pleasing to a flea, while the sound of a higher-pitched fork might be irritating. Various pitches are used to condition the flee to perform different actions.

In San Francisco, California, the Cardosa Flea Circus uses cat fleas *(Ctenocephalides felis)* rather than the human fleas (*Pulex irritans)* common in flea circuses years ago but rarely found today. The Acme Miniature Circus of Providence, Rhode Island, uses human fleas. Its stars, Midge and Madge, are shot out of a miniature cannon through a flaming hoop. The Alberti Flea Circus out of Winston-Salem, North Carolina, has performers named "Daring Diving Dardnell," "Captain Spaulding," and "Merlin." Some flea circuses have high wire acts and parades. Dogs are not welcome at the performances because the flea circus owners don't want to lose any of their performers. The next time you get bitten by a flea, be careful about swatting it. It just might be a star performer from a local flea circus.

FACTOIDS

A flea can pull 160,000 times its own weight, which is the equivalent of an average-size human pulling 12,000 tons.

Fleas can jump as high as 12 inches, which is 150 times their own length. If you could jump 150 times your height, you could jump about 1,000 feet in the air.

When a flea jumps into the air, it accelerates 50 times faster than the space shuttle after liftoff. It can also jump 30,000 times without stopping. A flea reverses its direction every time it jumps.

Swiss watchmakers used to make tiny circus vehicles for flea circuses.

The microscope used to be called the "flea glass," because the Dutch scientist Leuwenhoek, who helped develop it, used it to study fleas.

Only a few species of fleas bother humans, but over 250 species of fleas exist in North America.

DID YOU KNOW?

Many flea circuses lacked the one ingredient symbolic of the circus: clowns, or if you want to use circus lingo, "Joeys" or "Zanies."

Virtually all clowns fit into one of four categories: whiteface, auguste, hobo, or character.

A whiteface clown's basic white makeup covers his face but the other makeup, such as eyebrows, mouth, and cheeks, can be of any color or design. His costume tends to be more formal than that of the other clowns. Formal for a clown, that is. The costume colors are not as garish and clashing as those of the other clowns. This very serious whiteface clown is always the straight man in the circus. He takes charge of the action and sets up the routine. He is the one dishing out the jokes, such as throwing a pie or giving another clown a kick.

While the classic whiteface clown uses minimal makeup

around his facial features, the grotesque whiteface clown might have huge false eyelashes, a bulbous red clown nose, and makeup that makes his mouth look larger than normal. Ronald McDonald is in this category.

The auguste clown's base makeup is flesh colored, but the rest of the makeup is bright and exaggerates the nose, mouth, and cheeks. The clown's costume is garish, mismatched, bright, and usually oversized. Auguste was the personal name of the original auguste clown. He is the silly one who seems to never know what is going on. He is clumsy, awkward, impish, loves slapstick, and is usually the brunt of the jokes. He often works with a whiteface.

The hobo clown is a caricature of someone who has nothing and never will. He is a forlorn character who doesn't expect much out of life. The hobo clown is the one that gets the pie in the face or the kick in the pants. His costume consists of tattered and torn clothes, and his makeup exaggerates an unshaven face and large nose. Emmett Kelly and Red Skelton were both famous hobo clowns.

A character clown portrays some person or profession that the audience can relate to. Such characters might be policemen, firemen, doctors, cowboys, or sailors. Whatever the character is, the clown turns it into a caricature.

It doesn't matter if a clown is a whiteface, auguste, hobo, or character, his primary goal in life can be summed up in three simple words, "make 'em laugh."

More questions? Try these Web sites.

Famous Burials
http://interment.net/misc/famous.htm

If you want to know where famous people are buried, check out this site. It has a cemetery database of politicians, a "Find-a-grave" site for locating graves of famous people, burial places of the stars, and interment sites of the famous and infamous.

What Does Your Phone Number Spell?

http://www.phonespell.org/

This far-out fun site automatically tells you what word, or words, your telephone number spells. For example, the number 843-234-4663 spells "the beginner."

Paranormal

http://paranormal.about.com/culture/paranormal/?once=true&

This site is only for those who are interested in the paranormal, including such topics as the missing link, elves, antigravity, mysterious creatures, lost worlds, and ghosts.

The right side of the page ("In the spotlight") covers recent or popular events. You can click on any one of the 31 topics on the left side of the page to find a number of links with information about that topic.

Flea Circus

http://www.pe.net/~magical/flea/index.html

This is the Web site of Walt Noon, who not only has a performing flea circus but is also a professional magician. Click on the picture of Walt in a top hat to see a review of his circus. You can also hire Walt if you want a true flea circus at your next event. If you surf around his site, you'll also find some fascinating magic tricks.

6

Food

Why is salt a good food preservative? (Old salts are often well preserved.)

The purpose of "salting" meat is to preserve it so it can be eaten long after the animal has been slaughtered. In the presence of salt, water leaves animal tissues by osmosis, which dehydrates the meat so that no bacteria can grow and decompose it. There are a number of different salt preservation methods such as soaking in brine (a salt and water solution) or laying the meat in a bed of salt crystals.

In ancient times, salt was more highly prized than gold and was the foundation of ancient economies throughout history.

Aside from its materialistic value, salt also played an important part in religions of the world.

In the Christian religion, in both the Old and New Testaments, covenants were quite often sealed with salt. Jesus said that the meek, the poor in spirit, the merciful, those persecuted for the sake of righteousness, the peacemakers, and the pure of heart were "the salt of the earth," which was a marvelous compliment.

On the Sabbath, offerings in the Jewish temple included salt. Today Jews still dip bread in salt to commemorate those ancient sacrifices.

Many Buddhists believe that salt wards off evil spirits. When Buddhists enter their homes after attending a funeral, they throw salt over their shoulders to scare off any evil spirits that might be clinging to their backs. When the Dalai Lama died in 1933, he was buried sitting up in a bed of salt.

The Hopi Native Americans worship the Salt Mother and have a legend that the angry Warrior Twins put valuable salt deposits far from civilization to punish the human race. Only those who were brave and willing to work hard would be able to get the salt.

FACTOIDS

Salt was used as a form of money. Even today bars of salt are sometimes used in Ethiopia to pay for goods. Not too long ago cakes of salt with their value stamped on them were used in Tibet and Borneo. In ancient Rome a soldier's pay was salt, and our word "salary" for pay is derived from the Latin word *salarium,* whose root is *sal,* or "salt."

The word for salt is derived from the town Es-Salt, which is close to the Dead Sea.

The expression "not worth his salt" comes from ancient Greece, where salt was often traded for slaves. If a slave proved unworthy for some reason, it was said that he was not worth his salt. In short, he wasn't worth the price paid for him.

Many people throw salt over their left shoulder for luck if they have spilled some salt. This custom originated in ancient times when salt was so precious that spilling it was considered bad luck. To prevent misfortune, people would throw salt over their left shoulder into the eyes of the devil who was always dancing behind their left shoulder, hoping they would sin and he could have their souls. The salt would burn the devil's eyes and blind him until good luck returned.

DID YOU KNOW?

A salt mine in Wieliczka, Poland, is one of the top 12 tourist attractions in the world. It is the oldest salt mine in Europe and has been operating continuously for over 700 years. More than 200 miles of passages on nine levels extend 1,027 feet below the surface.

Through the centuries, miners have carved sculptures from the rock salt, and today there are underground churches with altars and life-size statues, all carved from salt. The ornate chandeliers are made from salt crystals. The Chapel of the Blessed Kinga, located 330 feet below the surface, is the largest church. It is over 164 feet long, 49 feet wide, and 39 feet high. Some of the large excavation chambers are used today as sports arenas or theaters. One chamber once served as an opera hall. Another fascinating attraction of the mine are the three underground salt lakes, one over 20 feet deep.

Today, around 1,500 miners are employed to work the mine. The miners are divided into gangs, and each gang works six hours before being relieved by another gang. Although people often use the expression "working in the salt mines" to mean working like a slave, the miners in the Wieliczka mine are quite healthy. They do not suffer from lung disease as hard rock and coal miners do. The temperature in the mine is about 54° F year-round. Each gang mines over 50 tons of salt a day, and the mine's annual production is around 75,000 tons of salt a year.

Although a great deal of salt is taken out of the mine every year, no one is worried about running out of salt. At the current rate of mining, it is estimated that it will take another 300 years to extract all of the salt in the mine.

Where was sourdough bread invented? (Was it the night they cremated Sam McGee?)

Although 6,000 years or so ago bread makers discovered that moistened flour fermented and expanded when exposed to air, sourdough bread as we know it today is associated with the prospectors who came to California during the Gold Rush of 1849 and later moved on to the gold fields of the Klondike.

To make a leavened bread, you need yeast. However, early prospectors did not have yeast, so they took advantage of natural airborne yeasts. These yeasts are living organisms that feed on dough and cause it to ferment. The by-products of fermentation are alcohol and carbon dioxide. The carbon dioxide gas makes the dough expand or rise.

Before the bread was baked, the cook would save a portion of the dough, which was already fermenting, and use it later to "start" a new batch of dough. The new batch was simply more flour and water that fed the yeast in the saved portion. This portion was called a "starter," because it started leavening a new batch of dough. So, to make proper sourdough bread, you must have a "starter."

At this point you might ask, "If I must have a starter before

I can make sourdough bread, how can I possibly make it if I don't have a starter to begin with?"

There are a couple of ways you can create a starter from scratch. One method is to simply put water and flour into a crock, cover the crock with cloth, and set it outside. Airborne yeasts will penetrate the cloth and start feeding on the flour and water mixture. In four or five days, the mixture will have fermented enough to use for making bread.

Another method is to make a paste of cooked potatoes, flour, and sugar, left outside until it is fermented.

When making sourdough bread, the general rule is "put in a cup and take out a cup." In other words, add a cup of flour to the starter for every cup you take out and use.

You can use the starter to make bread, pancakes, and rolls. If you take care of your starter properly, someday your grandchildren might be making sourdough bread from it.

FACTOIDS

When prospectors invaded California during the Gold Rush, their main diet consisted of beans, pork, and sourdough bread. Men who carried a crock of starter in their gear were called "sourdoughs."

Sourdough starters were treasures during the rugged frontier days, and some families handed down the starter through several generations.

The Hooch-in-noo Native Americans in southeastern Alaska took the liquid from the top of a batch of sourdough and let it complete its fermentation to produce a yellow alcohol. This potent mixture was said to cause a head-splitting hangover the following day. It was called "hooch"—a name still used today for homemade liquor.

Cooks on long trail drives kept their starter in five-gallon crocks. If it was a particularly cold night, they slept with their starter to keep the cold from stopping the fermentation.

Even today, some restaurants in San Francisco use sourdough starters alive since the California Gold Rush in 1849.

DID YOU KNOW?

The Canadian poet and novelist Robert Service lived and traveled through the Yukon for eight years. In 1907 he published a book of poems entitled *Songs of a Sourdough.*

One poem, "The Cremation of Sam McGee," gained international recognition. It told the story of a prospector who went to the Yukon in search of gold but found only frigid weather. Discouraged and cold, he felt he was about to die and made a friend promise to cremate his body.

When Sam McGee died, his friend put the body in the boiler furnace of a nearby abandoned ship. When the friend looked inside the furnace to check on the cremation, he saw Sam sitting in the flames smiling and saying it was the first time he had been warm since he arrived in the Yukon.

As Robert Service said in his poem:

The Northern Lights have seen queer sights,
But the queerest they ever did see
Was that night on the marge of Lake Lebarge
I cremated Sam McGee.

We can find pleasure in eating tasty sourdough bread, and we can also be pleasantly amused by the hundreds of tales about the old sourdough prospectors.

How are the hulls removed from sunflower seeds? (Chew and spit isn't a good method for commercial production.)

Removing hulls from sunflower seeds can be an elaborate process often requiring a variety of different machines. The hulls are typically removed with an air huller. Air is fed down the barrel of the machine at a high velocity. The high-speed air is propelled into a chamber containing the seeds. The force of the air hurls the sunflower seeds against a metal plate causing the husks to break free from the seed. Another machine is then used to separate the hulls from the seeds, while still another machine is used to clean the seeds.

If you enjoy eating sunflower seeds and can't afford an air huller, simply get a coffee can and fill it about half full with sunflower seeds. Punch a small hole in the can just large enough for an air hose to fit in. Then shoot the air into the coffee can and it will break most of the hulls and separate them from the seeds.

Many sunflower enthusiasts prefer the "chew and spit" method of separating the hulls from the seeds. When you use this method, you simply put the entire sunflower seed, along with its hull, into your mouth. You then chew the entire seed and spit out the hull when you're finished. It's a tasty but messy method.

"Chew and spit" requires some preparation, however. It's recommended that you cover the seeds (and hulls) with salted water, bring to a boil, and simmer for two hours. Drain them on a paper towel and then spread the seeds in a shallow pan. Roast in a 300°F oven for 30 to 40 minutes or until golden brown. Add one teaspoon of melted butter for each cup of seeds, stirring to coat them. Dry them on a paper towel, and salt to taste. Now all you have to do is find a wide open space and do your thing. Because these seeds can be very tasty, you'll definitely want to chew them. Just don't forget to spit.

FACTOIDS

When a sunflower plant is budding, it tends to follow the sun across the horizon.

Kansas is called the Sunflower State because the sunflower is so common there. In fact, the sunflower has actually become a serious weed problem.

Two types of sunflower seeds are grown commercially. The oil seed, high in oil content, is used to produce sunflower oil. It's also the seed preferred for most bird feeders. The non-oil seed, called confectionery sunflower, is larger, has black-and-white stripes, and is used in food products.

Although the sunflower is native to North America, Russia was the first country to produce it commercially.

Native Americans had many uses for the sunflower. They ground the seeds to make flour for mush, bread, and cakes. It was often mixed with other vegetables or eaten as a snack. A purple dye produced from the plant was used for textiles and other decorations. Other plant parts served as medicine. The stalks were even used as a building material. The sunflower was used as a hunting calendar: when the sunflowers were in bloom, buffalo were fat and the meat was tasty.

A single sunflower head can produce up to 1,000 seeds.

Russia produces more sunflower seeds than any other country, followed by Argentina and Eastern Europe. The United States is the fourth largest producer. We may not eat that many sunflower seeds but birds certainly do. The United States sunflower industry is estimated at $2.7 billion annually. Of that, about 15 percent, or $423 million, is spent on sunflower seeds to feed wild birds.

DID YOU KNOW?

More and more people are realizing that sunflower seeds are a healthy and fun snack. In fact, these tasty tidbits have even become a part of our national pastime, baseball.

Sunflower seeds now seem to be the snack of choice among fans and players alike. As one ballplayer remarked, "Seems like everyone's either eating them or playing with them."

In 1995, when the San Francisco Giants were playing against the Colorado Rockies on a brand new field in Colorado, the Giants outfielders weren't sure where to position themselves on the unfamiliar turf. The Giants manager said, "Just go where the seeds are." He meant that the Rockies outfielders were munching sunflower seeds and the telltale hulls would mark the best spot for the outfielders to situate themselves.

A pitcher who shall remain nameless loved eating sunflower seeds and kept digging them out of his pocket while he was pitching. Unfortunately, he stretched his fingers too far, injured them, and couldn't pitch for a while. He was the first known victim of SSF (Sunflower Seed Finger).

What is the difference between Cajun and Creole? (A matter of mispronunciation?)

Some 15 years before the Mayflower set sail for the New World, French settlers had already established a colony in Acadie, now called Nova Scotia, Canada. It was 150 years later when the English governor of Canada told the Acadians to forsake their Catholic faith and swear allegiance to England or be banished. They chose the latter and eventually settled in southern Louisana. Although called *les Acadiens* by the French, some referred to them as *le 'Cadiens*. As other settlers arrived in the area, they were not able to pronounce either term very well, and started calling them "Cajun."

The term "Creole" originally meant Africans born in the New World. No one knows for sure where the name came from. There are at least 30 different definitions, but one of the most popular is that the term came from the Portuguese word *crioulo,* which means "homegrown." The Creoles were influenced by African, French, and Spanish cultures, and today the term has come to mean the culture native to the southern part of Louisiana.

Both Cajuns and Creoles have contributed appetizing foods to our culture. For example, jambalaya is a Creole dish, while gumbo is a Cajun/Creole dish. Either can be made with a variety of ingredients. The easiest way to tell them apart is that gumbo is a soup and jambalaya is more like a casserole.

Gumbo is typically a soup that has meat or seafood and vegetables. Okra pods are used for thickening. The word comes from the Bantu word for okra, *kingombo*. If you've ever heard the song "Jambalaya," you've heard the term "filé gumbo." If gumbo contains ground sassafras as a thickener, it is called filé gumbo.

Jambalaya is typically rice cooked with a variety of ingredients such as tomatoes, onions, green peppers, herbs, and some kind of meat, poultry, or shellfish. There seems to be as many varieties of jambalaya as there are cooks. No one know for sure where the name came from, but the most common theory is that it is derived from the French word for ham, *jambon*. Although some people make a similar dish with noodles and call it jambalaya, it has to be made with rice to be authentic.

FACTOIDS

The 700,000 Acadians living in southern Louisana comprise the largest French-speaking minority in the United States.

Only Death Valley, California, is lower in elevation than the swamps that surround New Orleans.

The Cajuns have lived isolated from other cultures and

from each other. Thus, there is no single "Cajun" language. Numerous dialects are spoken, all derived from the original Acadian French.

DID YOU KNOW?

When people hear the word "jambalaya," many of them think of the song of the same name. It was written by Hank Williams, a superstar at 25 and dead at 29.

Hiram "Hank" Williams, Sr., was born in Georgianna, Alabama, in 1923. He never received any formal music instruction and learned to play the guitar from a black street singer named Rufus Payne, or "Tee-Tot." By the time he was 13, Hank Williams had a regular spot on WSFA, a local radio station. The station nicknamed him "the Singing Kid," and Williams stayed with the station for almost 10 years.

In 1949, he performed at the Grand Ole Opry singing his song "Lovesick Blues." The audience gave him a standing ovation and an unprecedented six encores, something not seen before or since. Within two years he was racking up hit after hit, including "Cold, Cold Heart" and "Hey, Good Lookin'."

In spite of his talent, Hank Williams had serious problems during his life. He went through a divorce in 1952 and was fired from the Grand Ole Opry a few months later. After hurting his back on a hunting trip, he took morphine for pain and became addicted. Even so, he released five hit songs that year, including "Your Cheatin' Heart" and "Jambalaya." Unfortunately, his life went downhill from there.

The following year he was scheduled to play a concert in Ohio. The weather was so bad that he decided not to fly and hired a chauffeur to drive him there. Hank Williams died in the backseat of his car while on the trip. His hands clutched a piece of paper with the words "We met, we lived, and dear we loved, then comes that fatal day, the love that felt so dear fades away." The autopsy report listed the cause of death as alcoholic cardiomyopathy.

The last record that was released before he died was "I'll Never Get Out of This World Alive."

What are the different types of caviar and why is caviar so expensive? (Roe, roe, roe your boat.)

Caviar is salted fish eggs, called roe, from the Turkish word *havyar,* meaning "bearing eggs." True caviar is made only from sturgeon roe. There are 26 species of sturgeon in the world, but only a few are commercially important as a source of caviar. In the United States, any fish eggs can be called caviar provided the name of the fish precedes the word caviar.

Many parts of the world produce caviar, but only the Caspian Sea has the unique combination of ideal water temperature, climate, and in-flowing rivers conducive to producing the finest caviar. In fact, almost 95 percent of the world's caviar comes from the Caspian Sea.

The sturgeons in the Caspian Sea are the beluga, osetra, and sevruga. Beluga caviar is the most prized and the most expensive. The eggs are larger than other caviar and their color ranges from dark steel to light gray. One reason for the cost is that at least 20 years must pass before a beluga sturgeon is mature enough to produce eggs.

Osetra eggs are smaller and golden yellow or brown. Although sevruga caviar is the cheapest of the three, it is also the most delicate. The eggs are small but have a strong flavor and are light to dark gray in color.

Prices of caviar vary according to market conditions. To give you an idea, beluga caviar sells for around $55 an ounce, while the least expensive sevruga caviar sells for about $27 an ounce. In some instances, demand has pushed the price of beluga caviar as high as $75 an ounce.

If you plan a party and decide to serve caviar, it's normally recommended that you plan on one ounce per person. If you only invite eight people besides yourself and your spouse, that means you would spend around $550 just for the beluga caviar.

Other countries produce inexpensive caviar from such fish as lumpfish, salmon, and whitefish.

With the overfishing taking place today and the increased demand for quality caviar, it's doubtful if the price will ever go down, but it's a good bet prices will rise in the future.

FACTOIDS

Caviar contains acetylcholine, which increases a person's tolerance to alcohol. Russians often drink caviar oil extract before drinking alcohol, believing it will prevent hangovers.

Up until the late 1800s, the United States produced 90 percent of the world's caviar from sturgeon in the Great Lakes. By then, overfishing had made the lake sturgeon virtually extinct and production of caviar was stopped. Today lake sturgeon caviar is only available in Canada. The Caspian Sea fisheries did not begin producing caviar until 1925.

It's rumored that when a wealthy Hollywood mogul had an affair with a young actress, his wife bathed in caviar because she thought it would keep her skin soft and young looking. She used her husband's credit card to pay for the caviar.

Sturgeon can live up to 100 years and grow to 20 feet in length.

Prior to Prohibition in the United States, there was such a large supply of caviar that New York bars gave it away to encourage people to drink more beer.

DID YOU KNOW?

Producing quality caviar is not a simple task. If the fish are caught in the open waters before they return upstream to spawn, the roe still has bite and texture. If they are caught in the tributaries just before they spawn, the roe tends to be soft and is likely to burst. Catching sturgeon is hard work. Some weigh up to 2,000 pounds. Some are hauled into the boat, while the heaviest ones are towed to shore.

The sturgeon is then taken to the fishing station. Everything that will touch the sturgeon or the caviar is sterilized. The fish is laid on a marble slab, cut open, and the roe carefully removed from its stomach and placed in a steel bowl. Because the ovaries are in a fine tissue, the roe is passed through a strainer to separate the roe from the ovaries and is salted and graded.

Grading is accomplished by lightly rocking the caviar back and forth in the steel bowl and inspecting the color, firmness, and size of the individual eggs. It is then packaged.

There is a great deal of labor involved in producing caviar, from catching the sturgeon to extracting and grading the roe. That is another reason for the high cost.

If you have a chance to eat some quality caviar, savor every mouthful, knowing the amount of work it took to get it from the Caspian Sea to your plate.

Why is coffee often referred to as "Joe"? (It's easier to say than Englebert.)

Although no one knows for sure, most people accept the naval legend. The U.S. Navy used to serve alcoholic drinks on its ships, typically wine. When Admiral Josephus "Joe" Daniels became secretary of the Navy, he made a number of reforms, including letting women into the Navy and abolishing wine from the officer's mess. Alcohol was outlawed on ships except for special occasions. Instead of drinking alcoholic beverages, the poor sailors had to drink coffee. Possibly out of sarcasm they referred to their coffee as a "cup of Joe."

Coffee is also called "java," a term coined by American hoboes in the late 19th century. It was derived from the coffee-producing country, Java.

Some authorities believe the word coffee comes from Caffa, an Abyssinian province. Others believe it's derived from the old Arabic word *qahwah,* which means wine. Coffee cherries were used to make wine long before the coffee bean was used to make coffee.

Whether you call it coffee, java, or joe, it's still one of the most popular drinks in the world.

FACTOIDS

Malays eat the leaves of the coffee plant because they contain more caffeine than the beans.

Caffeine is found in chocolate, some sodas (such as Mountain Dew, Coca-Cola, and Pepsi) but not others (root beer, ginger ale, Fresca), some pain relievers (Vanquish, Excedrin, and Midol) but not others (aspirin and Tylenol), as well as in some diuretics and cold remedies.

Coffee is the second most popular beverage in the world. Tea is the most popular.

In 1732, there was a movement to prevent women from drinking coffee because people thought it would make them sterile. Johann Sebastian Bach poked fun at the movement by composing his "Coffee Cantata," an ode to coffee. It included the aria "Ah! How sweet coffee tastes! Lovelier than a thousand kisses, sweeter far than muscatel wine! I must have my coffee."

The flowers of the coffee plant grow in large bouquets and have a faint odor of jasmine.

In 1906 an English chemist, George Constant Washington, was living in Guatemala. One day he noticed that a powdery condensation had formed on the spot of his silver coffee decanter. After numerous experiments with the substance, he created the first mass-produced instant coffee. (The first soluble instant coffee was invented by chemist Satori Kato of Chicago in 1901.)

A German coffee importer, Ludwig Roselius, took a batch of coffee beans that had been ruined and gave them to researchers. The researchers perfected a method of removing the caffeine from the beans while retaining the flavor of the coffee. The product was given the name "Sanka," and was introduced in the United States in 1923.

DID YOU KNOW?

The most expensive coffee in the world is Kopi Luwak, which sells for up to $300 a pound. What makes it so expensive?

A coffee bean is actually the seed of a small red fruit, called a cherry. Each cherry normally has two seeds, which we call coffee beans.

In Asia, there is a small carnivorous animal called the palm civet. It is also called the toddy cat because it has a fondness for palm juice, which the locals use to make a sweet liquor called toddy.

It is claimed that the palm civet eats the cherries whole but is unable to digest the seeds. While passing through the rodent's system, the seeds are partially digested by certain enzymes before they are excreted. Plantation workers collect the seeds and roast them to make Kopi Luwak coffee.

It makes a very good story and helps keep the price of Kopi Luwak coffee high. The coffee is even extolled by the Indonesia Tourism Promotion Board. However, it's not what it seems to be. The former head of the Indonesian national zoo said that the entire story is a fake, simply a great sales pitch for selling ordinary coffee at premium prices.

To date, no one has proved whether the story of the origin of Kopi Luwak is true, yet it is still being sold all over the world at premium prices.

It's sad that some people are paying up to $300 a pound for what might be just regular coffee. It's even stranger that they are willing to pay that amount of money for something they believe was first partially digested by an animal.

Perhaps it's all a matter of taste.

More questions? Try these Web sites.

Cheese

http://www.wgx.com/cheesenet/

Everything you ever wanted to know about cheese is on this Web site. Start by clicking on "Cheese index." You can then select any letter of the alphabet to list all cheeses with names beginning with that letter. It also includes a picture of the cheese and lists the country of origin.

The site also has a history of cheese, describes how cheese is made, and includes a cheese glossary. There is a section with stories and poems about cheese, as well as links to many other sites about cheese.

If for some reason you can't find an answer to your cheese question at this site, you can always ask Dr. Cheese. His e-mail address is: drcheese@wgx.com.

Adult Peanut Butter Lover's Club

http://www.peanutbutterlovers.com/

About three out of every four households in the United States consider peanut butter to be a staple like bread and milk. This Web site is a fan club for peanut butter lovers everywhere. It gives the history of peanut butter, explains how it's made, and contains other interesting information such as nutritional facts and recipes. There is also a trivia test you can take.

Fiery Foods

http://www.fiery-foods.com/

There seems to be a growing interest in hot foods such as chili, peppers, and other spicy foods. Click on "Cookin' with heat" to see numerous tips and techniques for spicy cooking. The three magazines on the site are slanted toward the producer or seller rather than the consumer, so you can skip those.

You should, however, click on "Hot links" for a great list of hot and spicy food sites. When you do, take a look at "Mark's remarkable hot links," which has just about everything you could want when it comes to fiery foods.

U.S. Food and Drug Administration (FDA)

http://www.fda.gov

The U.S. Food and Drug Administration (FDA) site covers not only food but also cosmetics, drugs, medical devices, and more. If you click on "Foods," you'll see a long list of related links. You can scroll down to find a topic that interests you, or you can search on a word by clicking on "Search" at the top of the page. A particularly interesting topic is "Food labeling and nutrition."

If you scroll down to the bottom of the home page, you'll see the "Special information for:" section, which has information specifically for consumers, women, and so on.

Junk Food Mecca

http://www.whpress.com/mecca/

This site lists most junk foods. Just click on the name of a junk food and you'll see the logo of the manufacturer. Click on the logo to go to that Web site. For example, if you go to the Moon Pie Web site, you'll learn the history of the Moon Pie, where the name came from, how they are made, and any collectibles and gifts you can buy.

7

The Human Body

Is it true that we use only 10 percent of our brain at any given time? (Some people use a lot less than that.)

Almost everyone has heard the statement "We only use 10 percent of our brain." Yet virtually no scientist in the world would agree with the statement, because it's an outright myth. We use all of our brain.

Why do so many people believe this myth? Probably because of the fascinating potential it offers. If we are only using 10 percent of our brain, then if we could learn to use the other 90 percent we might possibly perform super feats of memory, have superior intellect, control our bodily functions, and even move objects by just thinking about them. It's an appealing myth.

No one has been able to track down the origin of this myth. Some think it's a misquote of Albert Einstein. Others think it might have started because of a statement William James made in 1908: "We are making use of only a small part of our physical and mental resources." Still others think it might have been misconstrued from the valid statement "We only understand how 10 percent of our brain functions."

Let's examine the consequences of using only 10 percent of our brain at any time. First, why should such a large brain have evolved at all? Nature is efficient. If we are only using 90 percent of it, why make it so large? Scientists believe that all healthy brain cells participate in the brain's function.

Second, maladies such as Parkinson's disease and strokes have a devastating effect even though they damage only a tiny part of the brain. If we are using only a portion of our brain, then this small amount of brain damage should not cause a major problem.

FACTOIDS

The brain has roughly 100 billion brain cells.

There is no scientific evidence that older people can't learn new things. It may sometimes takes them a little longer, but they retain the new information as well as younger people.

Your brain has enough storage capacity to record 10 million books.

Growing old does not mean you will lose your memory. Current brain research indicates that if you keep your brain active, you will remain mentally sharp regardless of aging.

A signal from one brain neuron to another travels at about 200 mph.

Many scientists believe that the brain is the most complex structure in the known universe.

DID YOU KNOW?

A subject that often comes up in discussions of the brain is the phenomenal intelligence of "idiot savants." The phrase is a combination of the word "idiot" coupled with the French word *savant,* meaning "clever" or "learned." Today, the term has been replaced by the more appropriate "autistic savant."

A typical autistic savant may have an IQ of around 25, and may not be able to read or write, yet can perform amazing mental feats such as quoting the census of 5,000 U.S. cities, including the number of hotels in each city and the number of rooms in each hotel.

Some autistic savants excel in art. One drew such beautiful pictures that they were compared to Rembrandt's. Yet she lost her ability to draw once she learned to speak.

Another autistic savant was put in an institution because he could not care for himself. Although he had a very low IQ (50 to 80), he could easily translate among 16 different languages.

No one knows why autistic individuals, of whom there are numerous examples, have such incredible abilities. Perhaps one of the most fascinating examples is that of Blind Tom, known as the "Marvelous Musical Prodigy."

In 1862 Blind Tom seemed to be just a normal African-American boy of thirteen, except he was blind. He had never been educated in any way. In fact, until he was five or six years old he couldn't talk and could barely walk. Yet he had a capacity for music that few people possessed.

Blind Tom could play three musical pieces at once. While singing "Dixie," he would play the "Fisher's Hornpipe" on the piano with one hand while simultaneously playing "Yankee Doodle" with the other hand. He would also play many songs with his back to the piano and his hands inverted. His original compositions were called "picturesque, sublime, and a true embodiment of musical genius." In 1869 Mark Twain wrote an article about Blind Tom.

Perhaps we should not look down on those less intelligent than ourselves. It is quite possible that they may have a genius that we do not recognize or that they may offer the world something we cannot.

Is it true that a full moon makes people act strangely? (Lunatics or werewolves?)

It is a common belief that the full moon causes people to act strangely and that during a full moon there are more violence, suicides, accidents, aggression, and depression. The purported cause is sometimes called the "lunar effect" or the "Transylvania effect."

The words "lunacy" and "lunatic" come from the Latin word for moon, or *luna*. Based on this, some people argue that it has been known for centuries that a full moon has a strange effect on people.

In spite of these beliefs and other folklore and tradition, modern scientific studies have proven that there is no correlation between a full moon and unusual human behavior. Scientists have studied violence, crime, birth of infants, major disasters, kidnappings, alcoholism, sleep walking, depression, psychosis, suicides, hospital emergency room admissions, drug overdose cases, and accidents. In all of the studies, a full moon had no effect on any of these incidents. In fact, in many cases, such as suicide and drug overdose, there were fewer cases during a full moon than during a new moon.

If scientists have proven there is no correlation between a full moon and strange behavior, why do so many people still believe it is true? You often hear people say, "Just ask an emergency room nurse, or a bartender, or a police dispatcher and they'll tell you it's true."

Psychologists think that we believe in lunar madness because of folklore and tradition, the media, misconceptions, and selective memory. For example, we might remember some bizarre event that occurred during a full moon, yet forget that the same event also happened at other times.

We must also realize that if two events occur simultaneously, it doesn't mean that one event has caused the other. If it's raining outside and a football quarterback throws three touchdown passes in one game, it doesn't mean that the rain caused him to complete the passes, any more than making three touchdowns caused it to rain. Just because something happens during the full moon doesn't mean the full moon caused it.

FACTOIDS

There are countless legends about humans who changed into wolves during a full moon. These people were called "werewolves" from the Old English *wer,* meaning "man," and *wulf,* meaning "wolf." Do such beings exist?

There is a mental disorder called lycanthropy (derived from the Greek words *lycos,* meaning "wolf," and *anthropos,* meaning "man"). A person suffering from lycanthropy believes that he is a wolf or some other animal. The person's behavior may change until it resembles that of an animal, but the person is still just a human with a delusion.

Some individuals have a disease in which abnormal genetic traits cause a great amount of hair to grow all over their bodies.

The disease also causes dramatic receding of the gum line so that the person appears to have fangs.

Ergot, a parasitic fungus, often grows on rye and other cereal grasses. If rye flour contains ergot, it will be present in bread made from the flour. An outbreak of ergot poisoning occurred in France in 1955. People who were affected hallucinated and said they were "attacked by horrible beasts and terrified of the dark." Perhaps a similar outbreak in the Middle Ages fostered the werewolf legends.

All of the scientific evidence points to the fact that true werewolves don't exist. However, to be on the safe side, it wouldn't hurt to keep a wreath of garlic and a silver bullet by our beds. They will protect you against both werewolves and vampires.

DID YOU KNOW?

Was the moon once populated by furry, winged batlike humans? Today we'd laugh at such an idea, but in 1835 people believed it was true.

The eminent British astronomer Sir John Herschel traveled to South Africa to test a new telescope. In August 1835, the *New York Sun* ran a series of articles stating that Herschel had discovered life on the moon, including pygmy bison, beavers that walked on two legs, and the furry batlike humans.

By the time the fourth installment had appeared, the *New York Sun* boasted the largest circulation of any newspaper in the world.

It was all a hoax perpetrated by a British journalist who was trying to make a name for himself in the United States by boosting the circulation of the floundering *New York Sun*.

When Neil Armstrong walked on the moon, rumors circulated that the event was nothing more than a deception contrived by the U.S. government. Some time after the walk on the moon, the movie *Capricorn One* showed how such a hoax could be created.

They say that history repeats itself, but that's not always true. It's doubtful that furry batlike people walked on the moon, but it's a fact that Neil Armstrong did.

What makes your fingers and toes become pruney after you have been in the bathtub for a while? (Are you shrinking or expanding?)

A thick, tough layer of skin (in Latin, *stratum corneum*) covers the tips of your fingers, your toes, and the soles of your feet.

If you sit in a bathtub for a long period of time, or soak in a swimming pool or hot tub, your skin absorbs water and expands. Unlike the skin on the rest of your body, the skin on your fingers and toes has no place to expand, so it just buckles. This causes the skin to wrinkle, which gives it the "pruney" effect.

FACTOIDS

Not only did King Louis XIV of France never bathe but he never washed his feet either.

A typical drinking straw holds about one and a half teaspoons of water. To fill up the average 34-gallon bathtub, you would need the equivalent of 17,000 drinking straws.

Benjamin Franklin bathed regularly in a copper bathtub he imported from France. It was shaped like a shoe and hand-filled by a bucket.

In 1851 the first bathtub was installed in the White House.

During Prohibition, bootleggers made alcoholic drinks from alcohol, glycerin, and juniper juice. Because they used bottles or jugs that were too tall to be filled from a sink faucet, they filled the bottles under the bathtub faucet. Thus, the illegal concoction was dubbed "bathtub gin."

Queen Elizabeth I of England was thought to be overly meticulous, because she bathed as often as once a month.

In 1917 H. L. Mencken wrote an article explaining that the very first American bathtub was installed in the White House by President Millard Fillmore, who faced substantial public, medical, and legal opposition because of it. Although Mencken later admitted the whole story was a fake, it survived for almost 40 years. In fact, President Truman used the story in a speech in 1952.

Many states hold bathtub races. Contestants race in bathtubs equipped with outboard motors.

DID YOU KNOW?

Bathing was intended not only to keep people clean but also was involved in religious and magic rituals. Bathing was an ablution to remove the invisible stains acquired by touching the dead, committing crimes, or touching a diseased person.

Although homes in the Indus valley of Pakistan were equipped with bathrooms as early as 3000 B.C., the use of baths as a means of keeping clean was introduced much later by the ancient Romans.

Roman bath houses were public places that often included games, libraries, and stalls where goods were sold. The baths were warm, hot, and cold, and a Roman entered each one successively. The daily visit to the bath house was one of Rome's amusements.

Over time, people discovered that bathing could prevent disease, and it became a very private affair. In fact the bathroom today is a personal sanctuary where one can escape from the hubbub of life, other people, and even do some reading.

People in the United States consider a daily bath to be

essential. However, they usually take a shower to get clean and use the bathtub mainly for relaxation, which is not too different from the rituals used thousands of years ago.

Taking a shower can be invigorating and refreshing, but soaking in a bathtub full of warm water, surrounded by candles, can be a ritualistic event.

Why do your palms sweat when you are nervous? (I cannot tell a lie.)

You perspire to regulate the heat in your body (we're reminded of a grade school teacher who always insisted that pigs sweat but people perspire). If you become too hot, sweat glands in your body produce water droplets on the surface of your skin. As the water evaporates, it provides a cooling effect to your body. It's the same principle used by evaporative water air conditioners.

The majority of the roughly 5 million sweat glands in your body are concentrated on the palms of your hands, the soles of your feet and, to some extent, your armpits. About two-thirds of the glands are in your hands and are controlled by your sympathetic nervous system. So if it's hot and your hands, feet, and armpits sweat, this is normal.

However, if you are in a stressful situation and are nervous, angry, embarrassed, or anxious, your entire nervous system reacts and produces an immediate response called "flight or fight." Your nervous system has sensed danger and must prepare you for one of two options: combat the danger (fight) or run away from it (flight).

Your nervous system calls on all of your bodily functions to prepare you for this imminent danger. You start sweating, your heart beats faster, and you breathe more rapidly. This type of emotionally induced sweating is limited to your hands, feet, and armpits. Even if it's very cold, you will sweat if you are in a "flight or fight" situation.

Your palms may sweat for other reasons, such as an affliction that causes your system to produce more perspiration than what is needed to cool your body. Excessive and unnatural sweating can also be caused by hyperthyroidism, psychiatric disorders, menopause, and obesity.

We don't all perspire in the same way because the number and location of sweat glands are unique to each person. Climate also affects how we respond to temperature changes. If you grow up in a hot and humid area, you probably won't sweat that much. However, if you move from that location to a very dry area, you will probably perspire profusely.

It's good to remember that if your body doesn't need cooling and you are sweating, then you are in some type of stressful situation.

FACTOIDS

Although we use antiperspirants to get rid of body odor, perspiration by itself doesn't have a bad smell. The foul odor is caused by bacteria eating the debris on our skin.

Some people claim they never perspire. That's because their bodily thermostat is functioning efficiently and the perspiration evaporates the moment their body creates it. If perspiration is running off you, then your thermostat isn't working very well.

Because we have no control over our sympathetic nervous system, which controls perspiration, perspiration can be a factor in polygraph, or lie detector, tests. Along with heart rate, blood pressure, and respiration, perspiration gives a good clue as to whether the person is nervous or anxious and thus possibly lying.

DID YOU KNOW?

Perspiration, heart rate, blood pressure, and respiration are not the only bodily functions we cannot control. When we communicate with other people, especially in emotional situations, we may have conscious control over our words but not our body movements. These movements, referred to as "body language," are often more honest than what we say. Some authorities claim that body language accounts for 93 percent of our communication with others.

Suppose you are talking with a friend and you say, "Did you enjoy the ski trip with your boss?" He may say, "It was fine." However, if he doesn't smile and tightens his grip on the paper he's holding, you can probably bet that he didn't have a good time and doesn't want to talk about it right now. When people's body language contradicts their words, you're better off to go with the body language.

To give you an idea of what body language is about, here are a few examples:

Hands on hips: the person is either ready for something or showing aggression
Arms crossed on chest: the person is defensive
Rubbing an eye: the person doesn't believe what you are saying
Patting or fondling hair: the person lacks self-confidence

There are hundreds of examples of body language. The subject is detailed in many good books. If all the examples were listed here, you might rest your head in your hand and look downward. That would mean you're bored and that wouldn't be good.

Why do we have eyebrows? (The eyes have it.)

We have eyebrows for two reasons. The first is to keep water from running into your eyes. Your forehead can perspire more than other parts of the body. Perspiration is salty, and if you didn't have eyebrows it would run into your eyes and cause them to smart. If it is raining hard, water running off your head and down your forehead is stopped by the eyebrows so the water doesn't get into your eyes and hamper your vision.

You'll also notice that the bone under your eyebrows sticks out slightly. If you bump that bone, the eyebrows soften the blow to prevent damage to the bone. It is believed that early humans had much thicker eyebrows to provide more padding.

Your eyebrows, eyelashes, hair in your nose, and hair in your ears all serve to help guard your body cavities (eyes, nose, and ears) from insects and other foreign matter. Eyelashes, for example, keep dust, dirt, small insects, and other things out of your eyes when you blink.

The main value of hair is that it helps hold in body heat. That's one reason we have so much hair on top of our heads. A large percentage of body heat is lost through the top of the head. That's why years ago people used to wear "night caps" to bed so they would stay warmer.

Hair is also a sensory device. Hair grows out of tubelike pockets, called follicles, which have a number of nerves. That is one reason our skin is so sensitive when something touches it. Each follicle lies at an angle to the skin and when associated muscle fibers contract, the follicle is raised until it is almost perpendicular to the skin. This can cause what we call "goose bumps," or can cause our hair on the back of our neck "to stand on end."

Our hair is thus quite versatile, helping to keep us warm, protect parts of our body, and aid our sense of feeling.

FACTOIDS

At any given point in time, about 90 percent of your hair is growing while the remainder is just resting. A hair's growth cycle lasts anywhere between two and six years. It then rests for two or three months, then falls out, and a new hair starts growing in its place.

Blondes have the most hair on their heads, about 120,000 strands, while redheads have the least, about 80,000 strands or about a third less than blondes. People with brown and black hair have more hair than redheads but less hair than blondes.

Hair grows at different rates. It grows the fastest during the summer and when you are asleep. It also grows the fastest if you are between 16 and 24 years old.

A single human hair is stronger than a copper wire of the same thickness. In fact, the combined strength of all the hair on your head could support the weight of almost 100 people. In 1885, workers were having trouble rebuilding a temple in Kyoto, Japan, because the ropes were too weak to move the heavy building material and kept breaking. The temple worshippers all shaved their heads, and their hair was woven into ropes. The ropes were strong enough to move the heavy material and the temple was successfully rebuilt.

If you never cut your hair, it will grow to about 3.5 feet long. There have been a few people with hair more than 12 feet long.

A form of kung fu, called pak mei in the Cantonese language, means "white eyebrow," so called because the Buddhist monk who devised the system had silver eyebrows.

DID YOU KNOW?

If you've ever eaten sushi, then you already know a lot about human hair. Sushi rolls have three parts: a thick outer layer (usually edible seaweed), a center of rice stuffing, and a core of fish or vegetables.

Human hair is quite similar in structure. The outer layer, called the cuticle, helps protect the hair from damage. The center, called the cortex, is the main part of the hair. This contains the pigment that determines the color of the hair. The core, called the medulla, is the soft center of the hair.

There are many different theories to explain how hair grows. Yet in this modern age of science, the process that causes hair to grow and the reasons for hair loss remain a mystery. Until the mystery is unraveled, there will still be many bald men in the world.

Why does it feel good when you stretch? (Uncoiling the springs.)

Stretching sends a signal to the brain, telling it to make your muscles relax. As your muscles relax, you feel less tense. That's why it feels so good to stretch.

The muscles in your body are like springs. Imagine a spring that is very tight. You can't compress it much or produce much power with it. On the other hand, if a spring is very loose, it's quite easy to compress it. Once you let go of the compressed spring, it releases a great deal of force.

If your muscles are tight, they can't be contracted very far and they don't produce much power. Also, a tight muscle can't absorb much shock and puts even more strain on your joints. Tight muscles not only limit your performance in whatever you're doing; they can also lead to injuries.

If you spend just a few minutes a day stretching your chest, back, shoulders, and legs, you'll be more flexible and will feel a lot better.

Stretching may not be the fountain of youth, but regular stretching will keep you limber and make it easier to get in and out of your car or pick something up off the ground. You won't be any younger, but you'll feel younger. After all, youth is a state of mind.

FACTOIDS

Many people today have desk jobs and simply don't have the time or the desire to exercise. However you can stretch anywhere, anytime, in just a few minutes. Here are some stretches you can do while at work. Remember to breathe deeply, stretch slowly, and pay attention to the muscles you are stretching. You only need to do each exercise two or three times to feel better.

Stand up, place your hands on your hips, lean back as far as you can while looking at the ceiling, then lean forward to the upright position you started with.

Interlace your fingers and place your hands over your head with the palms facing up. Push toward the ceiling as you straighten your arms as far as you can.

Without moving your shoulders turn your head slowly to the left as far as you can, then to the right, and then back toward the ceiling.

Let your arms dangle at your sides. Rotate your shoulders by raising them up and back in a circular motion. Then rotate your shoulders forward.

Keep your feet flat on the floor, put your hands behind your head, and slowly arch your back while also bending your head backward. Don't do this if your office chair has casters or if the back of the chair can tilt, otherwise you may end up on the floor in a very embarrassing position.

Interlace your fingers, straighten your arms in front of you with your palms facing out, and push your arms out as far as you can.

DID YOU KNOW?

Stretching your body makes you feel good and has many benefits. Stretching your mind can do the same for you.

When you don't stretch, your muscles tighten up. If you don't exercise, your muscles get flabby. It's the same with the brain. It needs exercise and it needs to be stretched.

Scientists say that thinking creates dendrite connections in the neural pathways of the brain. That's a fancy way of saying that thinking creates a path in the brain for similar thoughts to follow. Over time, millions of these thought pathways are etched into your brain, and you tend to use the same paths over and over again when dealing with the many problems of living. It's as if your mind has become hard wired and reacts out of habit rather than thought.

The subjects taught in school tend to use the logical part of your brain, the left brain. Reading, writing, and arithmetic are best learned by the logical left brain. When you create art, play sports, or play music, you are using the creative right brain.

In day-to-day living, the left brain gets a great deal of exercise, but the right brain does not. If you create new pathways in your right brain, you will develop other ways of thinking, and you can also improve your insight. Insight is that wonderful feeling when the answer just pops into your head without your having to reason it out.

One of the best ways to exercise and stretch your entire brain is to solve puzzles, logic problems, and riddles. There are different kinds of thinking, such as reasoning in an orderly fashion (logical reasoning) or using analogies (analogical reasoning). Doing different types of puzzles will exercise all of the thinking processes of both sides of your brain.

To stay fit, exercise your body every day by doing stretching exercises. Don't forget to stretch your mind also by doing a crossword puzzle or trying to solve a riddle.

More questions? Try these Web sites.

HEALTH INFORMATION SITE

http://www.intelihealth.com/IH/ihtIH/WSIHW000/408/408.html

This site contains health information from Johns Hopkins University. It has health sections for men, women, children, and seniors. You can search for information on drugs, scroll through a medical dictionary, explore various diseases, and read a number of up-to-the-minute headline stories.

HEALTH CENTRAL

http://www.healthcentral.com/

One of the best medical sites on the Internet. It has late news, a section on fitness and weight loss, a library of over 6,000 diseases, free newsletters you can order, and a section hosted by the well-known Dr. Dean Edell.

The site also shows you how to start your own health profile and create a personal health page. It also has online doctor referral services.

THE HUMAN BODY

http://biology.about.com/education/biology/

Covers much of human biology, including altered memories, artificial corneas, brain cells, the fear center, neuronal fatigue, and taste buds. It also talks about such things as cancer and diet, how chocolate increases libido, why optimists live longer, and identical twins and fingerprints.

Inside the Human Body

http://www.imcpl.lib.in.us/nov_ind.htm

This site contains descriptions and drawings of the various systems within your body. Click on any system to display a drawing of that system and scientific facts about it. The seven systems covered are: circulatory, digestive, excretory, muscular, nervous, respiratory, and skeletal.

Brain Fitness—Exercise Your Brain

http://www.thirdage.com/cgi-bin/rd/hanl/living/games/brainfitness/

Your brain needs exercise as well as your body. This site has brain exercises ranging from meeting people to playing games. You can do most of these exercises at home or on your computer. Sections include brain warmups, brain aerobics, brain food, brain gymnastics, and brain energy.

8

Inventions

Who invented the zipper? (Have you ever heard of the "battle of the fly"?)

In 1893 Whitcomb Judson, a Chicago mechanical engineer, patented a device called the "clasp locker," which was the forerunner of the modern zipper. It had only two problems: it didn't work and no one wanted to buy it. Judson displayed it at the 1893 Chicago World's Fair. Although around 20 million people went to the fair, he sold only 20 of his new hookless fasteners. The U.S. Postal Service bought them to put on their mailbags.

Undaunted, Judson founded the Universal Fastener Company to manufacture his new device. One of the company's employees was Gideon Sundback, a Swedish immigrant who eventually became head designer.

Sundback made a number of improvements to Judson's original design, and by 1913 he had designed the zipper as we know it today. A patent was issued in 1917 for Sundback's design. It was called the "separable fastener."

When B. F. Goodrich ordered a huge quantity of them for the rubber galoshes he was manufacturing, he liked the "z-z-zip" sound they made and coined the name zipper.

At first, zippers were used mainly for boots and tobacco pouches. Almost 20 years later, in the 1930s, the fashion industry began promoting zippers for children's clothes. In the famous fashion "battle of the fly," the zipper beat the button hands down as French fashion designers started putting zippers in men's trousers.

Today zippers are found everywhere, on clothes, luggage, handbags, jumpsuits, and countless other products. In spite of competitive products such as snaps and Velcro, it appears that the zipper is here to stay.

If you look closely at a zipper, you'll probably see the letters YKK stamped on the pull tab. In 1934 a Japanese company called Yoshida Kogyo Kabushiki Kaisha, which means Yoshida Manufacturing Company, Limited, started manufacturing zippers. The company abbreviated its name to YKK, which became its trademark and was stamped on its zippers. Today, the YKK corporation has plants around the world. Its National Manufacturing Center in Macon, Georgia, is the largest manufacturer of zippers in the world. Since the center produces well over 1.5 billion zippers a year, there's a good chance that your zipper will have YKK stamped on it.

FACTOIDS

When the zipper was first introduced, people did not know how to use it, so directions were included with each zipper.

It's rumored that the zipper was invented because of a bad back. Supposedly a friend of Whitcomb Judson could not bend over to fasten his boots. Judson invented a slide fastener that his friend could open or close with one hand.

A Swiss mountaineer, George de Mestral, became very frustrated when walking through the woods because of the numerous burrs that stuck to his clothes. One day, while picking them off, he had an inspiration. He thought he could possibly use the same principle to invent a fastener that would compete with the zipper. He did just that and came up with Velcro. He liked the sound of "vel" in the French word *velours* (meaning "velvet") so he combined that with "cro" from the French word *crochet* (meaning "a hook"). Just two years after Mestral perfected his product, textile factories were churning out 60 million yards of Velcro a year.

DID YOU KNOW?

During the 1893 Chicago World's Fair, few people even noticed the inventor of the zipper, but virtually everyone acclaimed the genius of the inventor of the most sensational object at the fair, a gigantic vertical wheel that later became known as the Ferris wheel.

The wheel was 250 feet high with a circumference of 825 feet and weighed over 4,000 tons. It was powered by two 1,000-horsepower engines. The axle alone weighed 70 tons and was the largest piece of steel ever cast in one piece.

The 36 enclosed cars, resembling streetcars without wheels, could each seat 40 people, or a total of 1,440 passengers. If the passengers stood, the wheel could carry 2,000 people.

When George Ferris asked the fair directors for permission to build his wheel at the fair, they thought he was crazy. Fortunately for us, Ferris was a persistent man or today we would not be enjoying the numerous Ferris wheels in our country.

Oh, yes, the fair directors later said that George Ferris was a genius.

Why are soda cans cylindrical? (It all started with a bottle of wine.)

In 1795 the French wanted to find a new, efficient, and practical means of preserving food. Napoleon's armies had lengthy and vulnerable supply lines and food often spoiled before it reached the soldiers in the field. More soldiers were disabled by scurvy and hunger than by combat. A prize of 12,000 francs was offered to anyone who could come up with a solution.

An obscure Parisian, Nicolas Appert, had worked as a brewer and vintner, among other occupations. He had a brilliant idea: why not pack food in bottles just like wine?

It took him 15 years to perfect the idea and prove his theory that if partially cooked food was sealed in corked bottles and then immersed in boiling water, the food would not spoil. To test his method, he processed meat, vegetables, fruit, and milk. These were then put aboard French ships that set sail and stayed at sea for over four months. When the containers were opened, the food was as fresh and tasty as the day it was processed. Napoleon himself awarded the 12,000 francs to Appert.

Not to be outdone, the British decided to improve upon Appert's idea. Because glass could easily break, it could not endure harsh battlefield conditions. So they decided to make the containers out of metal. They designed a wrought-iron can lined with tin. There were only two problems with the can: it took a skilled craftsman a full day to make just four cans, and you could only open a can with a hammer and chisel. Eventually, a lighter can was made.

Cylindrical earthenware bottles were made in ancient Rome and in China. Later, bottles were made by blowing glass, which led to their round shape. Bottles were used in the first attempts at food preservation. Cans were a substitute for bottles, so they were made cylindrical like bottles. The custom has lasted until this day.

FACTOIDS

A modern can manufacturing line in a factory can turn out up to 400 cans per minute (almost 200,000 in an eight-hour day); a typical factory produces over a million cans a day.

Every three months, U.S. consumers throw away enough aluminum cans to provide all the aluminum required to completely rebuild all the commercial airliners in the country.

In spite of the popularity of cans, one study found that people preferred their soda in a glass bottle rather than a metal can.

Approximately 99 percent of all beer cans and 97 percent of all soft drink cans are made of aluminum.

From the time an aluminum can is made until it is recycled and remanufactured is an average of six weeks.

DID YOU KNOW?

For many people it's difficult to think of a tin can without thinking of soup. In fact, the term "soup can" has almost become a generic term for any can other than a soda or beer can. Soup cans are put to use even after the soup is gone. Boy Scouts and hoboes have used them for boiling water and for cooking. The cans have been used as tools for scooping and as banks to hold accumulated loose change.

Whenever someone thinks of a soup can, they almost always think of Campbell's condensed soup. In the late 1800s the Joseph Campbell Preserve Company developed a method for making commercially condensed soups. Because the water was removed from the soup, 32 ounces of soup were reduced to a mere 10 ounces and the price dropped from 34 cents to a dime.

Around the same time, a corporate executive from the company attended a football game between Cornell and Penn State and was so impressed with Cornell's brilliant red-and-white uniforms that he convinced the company to use the same colors on its soup labels. The red-and-white soup labels are still used today, almost a hundred years later.

The celebrated artist Andy Warhol is probably most famous for his painting of 200 Campbell's soup cans. When asked why he painted soup cans, he replied, "Because I used to drink soup. I used to have the same lunch every day for twenty years."

If it's good enough for Andy Warhol, then it should be good enough for everyone.

Who invented the bumper sticker? (Don't get stuck on this one.)

Shortly after World War II, fluorescent ink and self-sticking adhesive were developed. At that time a company called Gill-line, founded in 1934, specialized in printing and decorating canvas products. As far as is known, it was the first company to combine fluorescent ink and self-sticking adhesive to produce a sign that could be applied to an automobile's bumper. In effect, Gill-line invented the bumper sticker. When the company celebrated its fiftieth anniversary in 1984, it had already produced well over one billion bumper stickers.

You cannot drive very far these days without seeing a car with a bumper sticker. It's almost an American tradition. The stickers fall into many categories. There are sayings about work, such as "I'm a corporate executive, I keep things from happening." Some stickers refer to driving, such as "Cover me, I'm changing lanes," or "Honk if anything falls off my car." Still others can be satirical, such as those that mimic the movie *The Wizard of Oz,* "I owe, I owe, so off to work I go" and "Auntie Em, hate you, hate Kansas, taking the dog. Dorothy." And there are always the ones that are a play on words, such as "Horse lovers are stable people" and "Authors have novel ideas."

You know yourself how popular this product has become.

FACTOIDS

There are tens of thousands of bumper sticker sayings. You probably have your favorites. Here's just a few samples:

Ever stop to think, and forget to start again?
Change is inevitable, except from a vending machine.
Born free, taxed to death.
Forget about world peace . . . visualize using your turn signal.
As long as there are tests, there will be prayer in public schools.
I used to have a handle on life, but it broke.
A closed mouth gathers no feet.
A lost ounce of gold may be found, a lost moment of time never.
How in the world can buffalo fly with those little tiny wings?
It is bad luck to be superstitious!

DID YOU KNOW?

Bumper stickers let people express their feelings. But what did people do before the invention of the automobile? Well, it was simple. They inscribed the sayings on walls of buildings. Today we typically use the plural form of the old word for this kind of inscription, "graffiti."

Perhaps one of the earliest forms of graffiti is the "handwriting on the wall" mentioned in the Bible. The words *mene mene tekel upharsin* were written on the palace walls of King Belshazzar. Daniel said that these words meant:

> *mene:* God has numbered your kingdom and finished it.
>
> *tekel:* You have been weighed in the balance and found wanting.
>
> *upharsin:* Your kingdom has been divided and given to the Medes and Persians.

Although we think of graffiti as a modern phenomenon, it has been found on walls in Rome and at ancient Mayan sites in Mexico.

Today's graffiti is neither as prophetic nor as somber as that found in the Old Testament. Most of it falls into one of two categories: writing someone's name or organization, or a true attempt at an artistic design.

When people started painting their names and other sayings on walls, they were considered vandals who were defacing property and creating an eyesore. Their paint scribble might just be a name, a witty saying, or a declaration of love. These so-called "taggers" are still considered vandals today. In New York, graffiti was started in the early 1970s by an individual who wrote "Taki 183" on walls.

However, some people use graffiti as a serious art form and have been allowed to create elaborate colored graffiti in public places. These are called "pieces" (from the word "masterpiece") to separate them from the illegal taggers. These pieces are not uncommon in cities such as New York and Los Angeles.

In some cases, these graffiti pieces have become legitimate art. The work of the late Keith Haring, for example, moved from the subway walls of New York to the walls of respected art galleries and private art collectors.

Many pieces have been considered true art by some people, while others who have seen the identical pieces complain they are an eyesore.

Whether graffiti is good or just a nuisance seems to be in the eye of the beholder.

Who invented Gatorade? (It's definitely not alligator juice.)

During physical exertion at high temperatures, you can suffer severe dehydration and loss of body salts. If you lost just 2 percent of your body weight because of perspiration, your blood volume could drop, dangerously decreasing your blood pressure. Dehydration can cause fatigue, muscle cramping, decreased coordination, and poor all-around performance.

In the early 1960s, a team of researchers at the University of Florida started a project to develop a product that would rapidly replace the fluid and salts lost during extreme exertion. In 1965, Robert Cade, a kidney researcher and professor of medicine, concocted a formula to combat dehydration. All that was left to do was test the new formula.

Football players lose a great amount of body fluid when practicing and playing games. The researchers decided to test the new formula on 10 members of the University of Florida football team. The name of the team is the "Gators," derived from the word alligator, so it was decided to name the formula "Gatorade."

The football coach, recognizing the value of Gatorade, had his players drink it for the entire season. That year the Gators had a winning season and earned the nickname "the second-half team," because they outplayed all their opponents in the second half of the game. When the Gators whipped Georgia Tech in the 1967 Orange Bowl, the opposing coach said, "We didn't have Gatorade. That made the difference." *Sports Illustrated* magazine published the coach's remarks. From that time on, more and more football coaches realized that players needed to consume fluids during the game to avoid dehydration.

Robert Cade eventually sold the Gatorade rights to Stokely Van-Kamp, Inc. In 1983, the Quaker Oats Company purchased the rights to Gatorade. At that time, Gatorade came in three flavors, the original Lemon Lime, Orange, and Fruit Punch. There are now about 30 flavors worldwide. In spite of increased competition from about 60 different sports drinks, Gatorade is still the number one sports drink in the United States.

Whether football players are drinking it or pouring it over their coach's head in a victory celebration, the pale-green Gatorade will probably be part of the football scene for years to come.

FACTOIDS

To prevent dehydration when exercising, weigh yourself before and after your workout. If you have lost 2 or more pounds, drink 24 ounces of water for each pound you've lost.

In 1985, Jim Burt, who played for the New York Giants, was the first football player to dump a cooler of Gatorade on a coach when he drenched Bill Parcells. It wasn't until 1989 when a coach retaliated. Steve Spurrier, coach at the University of Florida, crept up behind three unsuspecting players who had doused him the week before and promptly dumped a cooler of Gatorade on them.

Drinking a sports drink is better than drinking water when you're working out. Water will quench your thirst but it isn't that good for preventing dehydration. It turns off your thirst before you get all the fluids you need, and it doesn't contain the important electrolytes that you lose when perspiring.

DID YOU KNOW?

When discussing Gatorade and the Gators, it's quite easy to think of alligators, which are very fascinating creatures. Their name comes from the Spanish *el lagarto,* meaning "the lizard." Although they are large, up to 16 feet long, and appear sluggish, they are extremely mobile and can roam as much as five miles a day, even across dry land.

It's not that difficult to tell the difference between an alligator and a crocodile. An alligator has a U-shaped snout rather than a pointed one, and you can't see its bottom teeth when its mouth is closed. That's just as well. An alligator has 80 large teeth and if any old ones fall out, new ones grow in. In fact, an alligator may go through as many as 2,000 or 3,000 teeth during its life.

When an alligator is depicted in a cartoon or on a shirt, it is always green for some reason. Real alligators are not green but black, so that the cold-blooded beast can better absorb the sun's rays and stay warm.

There are a few rare alligators that are albino, or completely white. Two rare white alligators are brothers, Boudreaux and Beauregard, owned by Jerry Savoie of Cut Off, Louisiana. He often lends them to zoos so the public can see these unusual creatures. Only 25 albino alligators are known to exist in the world.

It would be possible to write an entire book about these unusual creatures. However, this might be a good time to repeat an old saying, "See you later, alligator."

Who invented chewing gum? (Remember the Alamo!)

Humans have been chewing on something since the dawn of history, usually sap (resin) from various trees or wax. In 1848, John Curtis made the first gum in the United States when he cooked resin from a spruce tree on his wood-burning stove. In 1869, the first patent issued for chewing gum was given to William Semple, a dentist in Ohio, who invented a gum to exercise the jaws and stimulate the gums. It never sold, probably because it was made primarily of rubber.

The invention of gum, as we know it today, came about because of the friendship of two men, Thomas Adams, a photographer, and Antonio López de Santa Anna, who had defeated the Texans at the Alamo. When Santa Anna was exiled from Mexico, he lived with Adams on Staten Island, New York. Adams had tried different schemes to make money but all had failed. Santa Anna told him of an idea that could make Adams wealthy. He told him of a gummy substance that people in Mexico had been chewing for thousands of years. It was called chicle, the milky sap from the sapodilla tree that grows in the tropical rain forests of Central America. But gum was not on either's mind. The plan was to blend chicle and rubber together to make cheaper tires, toys, and rainboots.

Santa Anna had his friends in Mexico ship a ton of chicle to Adams. Although he labored for about a year, every one of his experiments failed. He had not been able to blend chicle and rubber. A vast amount of useless chicle was stored in his warehouse and Adams decided to throw it all into the river.

By sheer luck, Adams happened to go into a drugstore and saw a little girl buy some paraffin wax chewing gum. He remembered that Santa Anna had told him that Mexicans chewed chicle. Inspired, Adams started making unflavored pure chicle gum. It sold extremely well and Adams built a thriving business.

Some years later, John Colgan, a drugstore owner in Louisville, Kentucky, was selling a gum he made from balsam tree sap and flavored with powdered sugar. He had heard of how successful Thomas Adams was, so he ordered 100 pounds of chicle. He started making Taffy Tolu Chewing Gum, which was so successful that he sold his drugstore and devoted his time to manufacturing chewing gum.

A breakthrough in gum manufacture occurred when a popcorn salesman, William J. White, started experimenting with a barrel of chicle a friend had given him. He discovered how to flavor gum. Chicle does not absorb flavors, but sugar does. He combined flavors, such as peppermint, with corn syrup and then blended the mixture with the chicle.

In 1899, the major gum manufacturers united to become the American Chicle Company. William White was president and Thomas Adams, Jr., was chairman of the board. You might see some of their brands today, such as Black Jack and Beeman's.

People have been chewing gum ever since.

FACTOIDS

Clove gum became popular during Prohibition. Illegal liquor clubs passed it out as a breath freshener to remove traces of bootleg liquor on the breath.

Franklin V. Canning, a drugstore owner, invented the first pinkish-colored gum. He claimed it aided oral hygiene. He combined the words "dental" and "hygiene" to create a name for his new gum, "Dentyne."

The first bubble gum ever made, called "Blibber-Blubber," never sold well because it was too sticky.

William Wrigley, Jr., devised a marketing technique that caused sales of his gum to soar. He mailed one stick of gum to every person in the United States who was listed in the phone book.

DID YOU KNOW?

The world record for bubble-gum blowing is a 23-inch diameter bubble blown by Susan Montgomery Williams of Fresno, California. You may not be able to equal that feat, but professional bubble-gum blowers have some tips that will help your bubbles be bigger than ever.

1. The more gum, the better. Some people can hold up to 20 packs of gum in their mouth at one time. That's way too much for a beginner. However, if you have a lot of gum, you can blow bigger bubbles as long as you can keep moving the wad of gum around in your mouth.
2. Sugar doesn't stretch. Before you even start blowing, chew all the sugar out of the gum. Once the flavor is gone, you can start blowing bubbles.
3. Keep a smooth consistency. Chew the gum well to prevent air pockets from forming.
4. Stick out your tongue. Stick your tongue into the bubble as far as you can. That way you will get as much gum as possible into the bubble before you start blowing.
5. Keep your blowing uniform. Slow, steady blowing will produce a much more even, and larger, bubble.

With practice, these tips can help you blow bigger bubbles.

If during your practice the bubble should pop, wipe the gum from your eyes, go to the end of this chapter, look up the Web site on "Everything you ever wanted to know about bubble gum," and read the section entitled "Removing gum."

Good luck!

Who invented plastic? (Did it start with billiard balls or bugs?)

In the late 1800s, billiard balls were made of ivory. A United States inventor, John Wesley Hyatt, was trying to find a low-cost substitute for the expensive ivory balls. After many failures, he made a mixture of nitrocellulose, camphor, and alcohol, heated it so it could be molded, and let it harden. His discovery became known as "celluloid." Celluloid was strong, tough, and could be cheaply produced in a variety of colors. It was eventually used to make combs, toys, and other products.

Shellac, a secretion of an Asian beetle, was used as a varnish for preserving wood. It was also an excellent electrical insulator. In the early 1900s, electricity came into its own and there was more demand for shellac than could be produced by the cottage industry in Asia. A chemist, Leo Baekeland, began experimenting to produce a synthetic shellac.

After three years of experimenting, he heated phenol and formaldehyde to produce a liquid goo that when heated further turned into a hard but translucent substance. It was plastic and Baekeland called it "Bakelite." It wasn't long after its discovery in 1907 that Bakelite was used to make telephone handsets, radio cabinets, rosary beads, automobile distributor caps, cooking pot handles, and hundreds of other products. It is still used today in the manufacture of buttons, costume jewelry, pot handles, knife handles, and other items.

Celluloid was invented in 1869, almost 40 years before the invention of Bakelite. Yet Leo Baekeland is considered by most authorities to have been the inventor of plastic. The reason is simple. Celluloid is made from chemically treated cotton and other substances containing vegetable matter. Bakelite was produced by combining chemicals; no natural substances were used. As a result, it was the first 100 percent synthetic material.

FACTOIDS

The first production automobile to have an all fiberglass body was the 1953 Chevrolet Corvette.

The "House of Tomorrow" premiered at Disneyland in 1957. The walls, roof, floors, and furniture were all made of plastic. It was so strong that when it was torn down years later, wrecking crews had trouble demolishing it.

An experiment to create synthetic rubber failed, and the researchers ended up with a soft plastic they called "nutty putty." It was later sold as "Silly Putty," which became one of the most popular toys ever sold.

In the 1950s, many automobile manufacturers made entire cars of Plexiglas for display at auto shows. Even the engine was a clear plastic block so onlookers could see the internal engine parts. Unfortunately, curing the plastic with heat often created unwanted bubbles. Therefore, the plastic was cured at low temperatures by being put in a large refrigerator. It could take as long as six months to cure the plastic this way. Although a plastic Coke bottle was first produced in 1958, it didn't meet commercial standards. It took almost 20 years more to develop an acceptable plastic commercial Coke bottle.

DID YOU KNOW?

People today don't pay much attention to cellophane, but at one time it was so valuable that it was locked in safes at night.

It all started around the time Bakelite was invented. A Swiss chemist, Jacques Brandenberger, was tired of looking at unwashed and stained tablecloths in Paris restaurants, so he decided to invent some type of covering that could simply be wiped clean. He sprinkled liquid viscose on a tablecloth and it became glossy but very stiff. He then decided to try to make some type of transparent sheet that could be laid over the tablecloth.

By using a cellulose product, he managed to create a transparent sheet. He combined the French words *cellulose* with *diaphane,* meaning "transparent," and called his invention cellophane.

Brandenberger tried selling cellophane to the movie industry to replace the highly flammable film they were using at the time. Unfortunately, cellophane became distorted when hot, so he decided to sell it as a wrapping for expensive products. The cost of cellophane was so high at the time that only the most exclusive Paris boutiques could afford to wrap their finest perfumes in it. At night, they kept it locked in the store safe.

Over time, cellophane became less expensive, and eventually DuPont scientists created a version that was moisture proof. From that time on, cellophane became the packaging of choice for candy, cigarettes, and a host of other products.

The next time you receive a gift wrapped in colored cellophane, just be thankful you don't have to lock it up in a safe at night.

More questions? Try these Web sites.

INVENTORS AND THEIR INVENTIONS

http://inventors.about.com/education/inventors/msub12.htm

This wonderful site lists most major inventions. Just click on a letter of the alphabet to find a list of inventions beginning with that letter. Once you find an invention you like, click on it to see a Web page full of links related to that invention. The site covers most inventions from air bags to the zipper.

If you'd prefer to look up the inventor, the following page lists inventors from Edward Acheson (who invented carborundum) to Vladimir Zworykin (who invented the cathode-ray tube): http://inventors.about.com/education/inventors/msub1_1.htm

ALL ABOUT GUM

http://www.nacgm.org/index.html

The National Association of Chewing Gum Manufacturers has an excellent site. Click on "Consumer section," and you'll find topics such as "The story of gum," "How gum is made," and even "Tips for getting unstuck."

If you enjoy bubble gum, you should have a lot of fun at Pud's Clubhouse at: http://www.dubblebubble.com/clubhouse/index.html

EVERYTHING YOU EVER WANTED TO KNOW ABOUT BUBBLE GUM

http://www.welcome.to/BubbleGum

Ms. Demeanor's bubble gum page is chock full of information about bubble gum, including history, fun facts, ingredients, making gum, gum music, languages, links, fun and games, books, removing gum, bubble blowing, cocktails, gum forum, and gum chat, frequently asked questions, gum e-mail, and much more.

Make sure you scroll down the left side of the page to see all of the topics.

Gallery of Obscure Patents

http://www.delphion.com/gallery

This fun site has drawings and descriptions of real, although strange, patents. Some of the inventions shown include a motorized ice cream cone so you don't have to turn it while licking it, a Santa Claus detector, a gravity-powered air conditioner for shoes, a fly swatter that plays music, an enclosure to protect you from killer bees, and a greenhouse helmet that contains plants so you can breathe the oxygen they give off.

Another site with weird inventions is http://www.lightlink.com/bbm/weird.html

Greatest Inventor of All Time

http://www.mos.org/sln/Leonardo/

Some of the most fascinating inventions of Leonardo da Vinci are shown here. Click on "Inventor's workshop" to see some of the inventions. Once there, you'll find other links. For example, if you click on "Visions of the future," you'll see modern machines that Leonardo described and sketched, including the parachute, helicopter, tank, and scuba gear.

If you want to have some fun, click on "Leonardo's mysterious machines." Click on one of the numbers to the left to see a drawing of the machine. Then see if you can guess which of the four machines it is by clicking on the appropriate word beneath the picture.

9

Language

Where did the term "You've been 86'ed" come from? (Throw it out the back door.)

Many years ago, Chumley's Restaurant, at 86 Bedford Street in New York City, had a custom of throwing rowdy customers out the back door. Eventually, restaurant workers started using "86'ed" as a synonym for something being thrown out.

The term soon became part of the colorful "hash house" or "lunch house" jargon adopted by the brash and often sassy waitresses and countermen who worked in diners in the 1930s, 1940s, and early 1950s.

In those years it was a common practice for a waitress to call out the order to the cook. For more efficient ordering, the waitresses created a shorthand method of giving orders. The jargon they developed used just a few very descriptive words to make sure the cook understood the order.

For example, if you ordered two poached eggs on toast and a glass of milk, the waitress would yell out to the cook, "Adam and Eve on a raft and moo juice." If you ordered a toasted English muffin with coffee, the waitress might yell, "Burn the British and Joe."

Although the colorful language of the diners died out many years ago, a few words have remained in use today, such as, O.J. (orange juice), stack (pancakes), BLT (bacon, lettuce, and tomato sandwich), mayo (mayonnaise), and the various ways of cooking eggs such as over easy or sunny side up.

Years back it was fun to eat at the lunch counter and try to figure out what the waitress was saying. It's rare to hear this colorful slang today but you may still hear it in some of the old lunch counters in the smaller, older cities of the United States. There are also a few diners that use nostalgia as their theme and thus use the old hash house lingo.

FACTOIDS

Here are just a few examples of hash house lingo:

Axle grease: butter
Blowout patches: pancakes
Bowwow, barks, or groundhog: hot dog
Baled hay: shredded wheat
Bessie: roast beef
Bessie in a bowl: stew
Cackleberries: eggs

Rabbit food: salad
Sinkers: dougnuts
Sweep the kitchen: hash
Whiskey: rye bread (derived from rye whiskey)
Yummy: sugar

DID YOU KNOW?

When reminiscing about the old diners, you may think of one of the most popular meals they served: the hamburger.

Although it is widely accepted that the name was derived from the town of Hamburg, Germany, around 1884, people still argue about who created the hamburger as we know it today. Some insist it was Charlie Nagree, who sold burgers at a county fair in Wisconsin in 1885; while others claim it was Louis Lassen, who featured them at his lunch wagon in New Haven, Connecticut, in 1900. There are many other claims.

However, there can be no argument that one of the first fast-food restaurants to feature hamburgers was White Castle. With a loan of $700, the first White Castle establishment opened in Wichita, Kansas, in 1921. It charged five cents for a hamburger.

In 1930 White Castle hired a well-known food scientist to determine the nutritional value of White Castle hamburgers. During one experiment, a college student lived on nothing but White Castle hamburgers and water for 13 weeks and remained in good health. People have been eating them ever since.

Today most people would probably admit that the king of fast-food hamburger stores is McDonald's. In 1955 McDonald's had one establishment in Des Plaines, Illinois. Today the company has over 24,500 stores in 116 countries.

Yet it seems that our appetite for hamburgers has not diminished in the least, as evidenced by the growing number of chains such as Burger King, Wendy's, In and Out, and many others.

If you were to order a hamburger in a hash house diner, the waitress might say, "Bossy on a raft and hold the mayo."

How many geese are in a gaggle? (If you poke a goose, are you giving it a people?)

The word "gaggle" simply means a group of something. It does not denote any specific amount. "Herd" is a good example of a similar type of word. We say a herd of cows. The herd could be 20 cows or 2,000 cows. Either way, it's still a herd of cows.

It seems that almost every bird and animal has a word to denote a "group" of them. Here are a few examples:

Army of frogs
Bale of turtles
Band of coyotes
Bevy of quail
Bouquet of pheasants
Cast of hawks
Charm of finches
Clowder of cats
Colony of penguins
Company of plovers
Crash of rhinos
Descent of woodpeckers
Dole of doves
Down of hares
Dray of squirrels
Drift of hogs
Float of crocodiles
Gang of elks
Kettle of hawks (no, not a "kettle of fish")

Kine of cows
Labor of moles
Leap of leopards
Murder of crows
Peep of chickens
Shrewdness of apes
Sleuth of bears
Wedge of swans

FACTOIDS

If a goose gets sick or wounded while in flight, two other geese drop out of the flying formation and escort the stricken goose to protect it. They stay with it until it either dies or is able to fly again.

Although geese are very social animals among their own kind, they have a tendency to bite people.

When migrating, a snow goose flies at speeds up to 50 mph and will travel almost 3,000 miles at an altitude of almost 3,000 feet.

Geese mate for life and grieve at the loss of a mate.

The most widespread goose in North America is the Canadian goose. Unlike most birds, the family stays together after the breeding season.

Small birds sometimes hitchhike on the backs of flying Canadian geese.

DID YOU KNOW?

Perhaps the most famous goose is not a goose at all. It's an airplane called the Spruce Goose.

During World War II, enemy submarines were devastating U.S. merchant ships. The cargo ships were being sunk as fast as they could be built. Henry Kaiser of shipbuilding fame thought that a solution would be to create a cargo plane that could fly over the water and deliver needed materials to the troops.

Kaiser sought the aid of legendary airplane designer Howard Hughes and the two formed a consortium to build a giant transport plane. It was designated the HK-1 (for Hughes and Kaiser) and was originally called the Hercules.

To avoid using critical war matériel, they decided to build the plane from wood, not realizing at the time how difficult it would be to make an airframe of birch and spruce.

Although Kaiser dropped out of the project, Hughes succeeded, and his flying boat was the biggest airplane ever built at that time. Because of its wooden body, it was nicknamed the Spruce Goose.

The plane had a wing span slightly longer than a football field (320 feet), and was over 218 feet long and almost 80 feet high. Its eight 28-cylinder Pratt & Whitney engines had 17-foot propellers. Designed to carry 65 tons, it was capable of carrying 750 fully equipped troops or two Sherman tanks.

Unfortunately, it was not completed until after the war ended, and the only time it ever flew was an unauthorized flight by Howard Hughes, who flew it about a mile at a height of 70 feet to prove it was airworthy.

It was kept in Long Beach, California, for some time. The Walt Disney Company bought it but decided to get rid of it after a few years. It was bought by Evergreen International Aviation. They had it disassembled and then had each section shrink wrapped. The sections were loaded onto an ocean barge, which took it to Portland, Oregon. It was then shipped up the Willamette River to the site where its overland journey began. Three pieces of heavy moving equipment, each with 104 forward gears, moved it to its final location. Today, the famous Spruce Goose is at the Evergreen Aviation Education Center in McMinnville, Oregon.

Although there were plans to build more of the Hughes flying boats, only one was ever built. This is one goose that will never be part of a gaggle.

What is the origin of the word "quack," meaning a fake or unethical doctor? (Duck if you run into one of these doctors.)

The word "quack" is an abbreviation of "quacksalver," a 16th-century word meaning a peddler who sold fraudulent medicines in the street. Since to quack meant to peddle, to "quacksalve" meant to peddle a salve or ointment, especially one that allegedly could cure any malady. The charlatan who did this peddling was a quacksalver. Today a quack means either a fake or unethical medical doctor.

Although medical quackery has been around for centuries, it seems to have flowered the most in the United States. The medicine wagons of the snake oil peddlers in the 1800s later gave way to much more sophisticated scams.

Radionics was one of the most popular frauds in the 1920s and 1930s. The quack doctor used a small wooden box with a number of holes in the front and a light bulb inside. A number of wires, a dial, a pedal, and a glass tube were added to make the box impressive. Thin pieces of colored paper were pasted over the holes.

The patient would moisten a slip of paper with his tongue and then drop it into a slot on top of the box. The quack doctor claimed he could see a letter through one of the holes that would diagnose the patient's illness. For example, an *A* meant the patient was suffering from arthritis. Radionic doctors claimed they could lengthen a patient's legs, cause amputated fingers to grow back, and fill dental cavities, as well as kill dandelions over any specified distance and fertilize fields as far as 70 miles away.

Another fraud was the Spectro-Chrome therapy machine. The quack doctor explained that each color had significance. For instance, red energized the liver, blue built vitality, lemon yellow built bones, purple prevented malaria, and so on. When these colors were not in harmony, disease or other physical ailments resulted.

The quack doctor claimed his Spectro-Chrome machine would attune all the color waves of a person's body to bring good health. As bizarre as it sounds, by 1940 he had sold enough of his machines to earn him over $1 million.

There are countless other medical frauds in our history, most of which were invented in the late 1800s or early 1900s. A few examples are the Homo-Vibra Ray, whose practitioners claimed the ability of diagnosing illnesses even if the patient was a great distance away; the electric brush designed to relieve headaches and promote hair growth; the galvanic eyeglasses said to improve eyesight and prevent nasal congestion; and the radio disease killer, which was an impressive-looking box of electronic gear that the patient could use to cure himself.

Although it may seem odd that such fake devices could ever sell, when people are ill or suffering and cannot find a cure, they will often try anything.

Today there are many reputable alternatives to standard medical techniques, but there are also many quacks out there.

FACTOIDS

Individuals frequently made medicines at home, often containing opium or alcohol, which they claimed could cure most ailments. The inventor would then seek a patent for his concoction. Hence the name patent medicine.

A typical patent medicine might claim it could cure sour stomach, headache, nausea, coughs, cold, consumption, pneumonia, asthma, bronchitis, and pleurisy.

One man made a very successful machine called the "Magno-Electric Vitalizer," which he claimed could cure almost anything, including rheumatism, deafness, and paralysis. When his company was shut down for fraud in 1904, his father, the well-known inventor Thomas Alva Edison, said, "My son has never shown any ability as an inventor or electrical expert and is incapable of making any invention or discovery of merit."

DID YOU KNOW?

Will Keith Kellogg, the creator of the Kellogg cereal company in 1894, once worked as a bookkeeper and manager at the Battle Creek Sanitarium, where his older brother, John, was a staff physician.

John Kellogg's surgical skills were admired by well-known doctors of his time, but he began using questionable medical practices and later was called a quack.

During this time both Will and John Kellogg searched for a digestible substitute for bread. They experimented by boiling wheat. They never did find a substitute, but by accident Will came up with something far better.

Will had forgotten about a pot of wheat he had boiled, and it softened and congealed after standing for a long time. When he finally put the wheat through the normal rolling process, each grain emerged as a large, thin flake.

He decided to serve this to the patients. They liked the wheat flakes so much that it eventually became one of their favorite foods.

After leaving the hospital, many patients wrote Will and asked for packages of the wheat flakes. He began packaging the flakes and his small enterprise eventually grew to become the Kelloggs Company.

What is the origin of the word "cocktail" for a mixed drink? (A cocktail can make you cockeyed.)

No one really knows for sure. There are at least 14 explanations of the origin of the word and authorities do not agree on any one of them as being factual. Here are a few of the more common explanations.

When preparing roosters for a cock fight, trainers often made a bread from a mixture of flour, stale beer, white wine, gin, seeds, and herbs. They fed this to the birds to ready them for the fight. This bread was called "cock-bread ale" which was later shortened to "cock-ale." This eventually became "cocktail."

In the 17th and 18th centuries, beer was mixed with minced meat of boiled cock and other ingredients and was called "cock-ale," which evolved into "cocktail."

In 1926 a French writer claimed that the word *cocktail* was derived from the French word *coquetel,* which referred to a mixed drink produced in the Bordeaux region of France.

Another source claims that the word is short for "cock tailings," which are the remains of various liquors all put into one container and sold cheaply.

FACTOIDS

The word "punch" comes from a Hindi word for "five." Perhaps coincidentally, the English colonists in India often used five ingredients when making the drink: rum, tea, sugar, lemon, and water.

Bourbon derives its name from Bourbon County, Kentucky, where it was first produced. However, Bourbon County no longer produces bourbon. Although some counties in Kentucky produce bourbon, it is illegal to buy it in these counties.

Trader Vic (Victor Bergeron) created a new drink and served it to his friends from Tahiti. After tasting it, they exclaimed "*Mai tai—roa ae!*"—"out of this world—the best" in Tahitian. Mai tai became the name of the drink.

When the pilgrims loaded the Mayflower before sailing to the New World, they stored more beer than water.

DID YOU KNOW?

There is a mystique about one of the most famous cocktails of our time, the martini, made with gin and dry vermouth. The proportions depend on the type of martini:

Traditional: two parts gin to one part dry vermouth
Dry: five parts gin to one part dry vermouth
Extra dry: eight parts gin to one part dry vermouth

There are countless stories about martinis and famous people such as President Franklin Roosevelt, who added fruit juice or a teaspoon of olive brine; W. C. Fields, who called double martinis "angels' milk"; and Ernest Hemingway, who called his martini "the Montgomery," because World War II English Field Marshal Montgomery liked battlefield odds of 15 to 1 and Hemingway liked martinis with 15 parts of gin to 1 part of dry vermouth. There are also famous quips such as the one by Robert Benchley. He had spent most of the day floating in a studio water tank. When he emerged he allegedly said, "I must get out of these wet clothes and into a dry martini." (Yes, it was said by Robert Benchley, not by Alexander Woolcott, Charles Butterworth, or Mae West.)

Probably the most famous saying about martinis is the one that Agent 007, or James Bond, uses when he orders one. "Make it large and very strong and very well made." After a short pause he continues, "shaken, not stirred."

Could James Bond have possibly been wrong? Most bartending books say that a martini should be stirred. A good cocktail should be served cold, clear, and not diluted with water. Shaking a drink chills it better than stirring it. However, shaking it traps air bubbles and clouds the drink.

Some claim that more ice melts when a drink is stirred, while others claim more ice melts when it is shaken.

Well, it appears to be a toss-up, so it's not possible to determine if Agent 007 is correct or not. The controversy of shaken versus stirred will never be resolved. There are staunch proponents on both sides.

Unless you happen to be employed as a government agent, why not just enjoy the James Bond films the way they are and drink martinis made the way you personally prefer?

What is the origin of the word "shrink" when used as a name for a psychologist? (Don't lose your head over this one.)

The term "shrink" is a shortened form of "head shrinker," which was not meant to be a compliment. Many cultures throughout the world were headhunters who took the heads of their enemies as battle trophies. However, the only known practice of shrinking heads occurs in remote parts of Ecuador, Peru, and Brazil. The Jivaros, who live in the Amazon jungle, gained international attention when they were first discovered. They practice a ritual of drying and shrinking the heads of their slain enemies.

Jivaros are feared throughout South America. In fact, they

are one of the only tribes who have successfully defended their homeland against Spanish invaders. In one secretly planned attack, the Jivaros massacred 25,000 Spanish settlers. The Spaniards made numerous attempts to conquer the Jivaros, but never succeeded. Although their land has rich gold deposits, their ferocity has discouraged outsiders from entering their territory.

Jivaros believe that a shrunken head holds both the power and the soul of the victim. They also are convinced that when a person is killed, the soul seeks revenge. By shrinking the head, the warrior not only gains the power of the victim but also prevents the avenging soul from harming anyone because it is now trapped in the shrunken head.

To the headhunters, shrinking heads was both a religious and magical practice. The term "head shrinker" or "shrink" implied that psychologists were also using magic, rather than science, in an attempt to cure their patients. Although it was a term of ridicule when it first appeared in the early 1960s, it has since become a common term. It no longer implies contempt and is often used in a humorous sense.

FACTOIDS

Of all the people in this country who are between the ages of 15 and 54, about half of them have had a psychological problem during their lifetime.

The results of a study done by a large corporation showed that 60 percent of employee absences were caused by stress, a psychological problem.

Six facial expressions are universally recognized: fear, anger, surprise, disgust, sadness, and happiness.

Between 50 and 70 percent of patient visits to a primary

care physician have nothing to do with a physical illness but are related to psychological factors. Depression and anxiety are among the six most common conditions seen in family practice.

DID YOU KNOW?

Today the Jivaro tribe of southern Ecuador still performs the ritual of shrinking the heads of slain enemies. It takes quite a bit of time to shrink a head. The Jivaro warrior first makes an incision in the back of the head in order to remove the skull. He then turns the skin inside out and scrapes it. Later he sews the eyes and lips shut.

He boils the head in a mixture of herbs and water for about two hours, which causes the head to begin shrinking. At this point, it is about one-third normal size. Once it has cooled, he sews up the incision at the back of the neck and successively fills it with hot rocks and sand which continue the shrinking process until the head is about the size of a baseball or a man's fist. He then hangs it over a smoldering fire all night which changes the color from yellowish to black. As the final step, he rubs the head with charcoal dust to polish it.

About 150 years ago, there was a lively trade in shrunken human heads. The heads were sought by museums, collectors, and curiosity seekers. Eventually both the Peruvian and Ecuadorian governments passed laws that dispensed severe penalties for anyone buying and selling shrunken human heads. However, there was still a great demand for the shrunken heads, so people started making realistic fakes to sell. These counterfeit heads were carefully crafted from animal skins and often fooled even museums. Even today these fakes are sold as souvenirs and the older, classical fakes can fetch a very high price, often $1,000 or more.

People seem to be fascinated by shrunken heads and often

buy a replica made out of rubber at a fair or novelty store. These can sometimes be seen hanging from rearview mirrors in automobiles.

Perhaps a less gruesome and more appealing ornament to hang from a rearview mirror would be fuzzy dice.

Where did the term "chickenpox" come from? (What about cowpox and monkeypox?)

The term "chickenpox" originated in 1720. It is the popular name for varicella, a mild eruptive disease that bears some resemblance to smallpox. It chiefly affects children.

The name is based on the mildness of the disease compared to the much worse and often deadly smallpox. The chicken is considered the mildest of all barnyard fowl (that's one of the reasons we use the word chicken as a synonym for cowardly).

The term "pox" actually comes from smallpox rather than chickenpox. Persons infected with smallpox first break out in a red rash that eventually turns into blisters. If the blisters are scratched or broken for some reason, they leave a "pock" mark or scar on the skin. The person can thus become covered with "pocks," from which the word pox is derived.

Most children in the United States come down with chickenpox before they are 10 years old. Although chickenpox is typically not dangerous, recent studies have shown that aspirin should not be given to a child that has the disease. Doing so could cause Reye's syndrome, a rare brain malady.

People who have had chickenpox and are cured are immune to the disease for the rest of their lives.

Most people have heard of smallpox and chickenpox, but not many are familiar with cowpox. Cows carry a virus very similar to smallpox that causes a very mild disease. Many people living in the country knew that if you had been sick with cowpox, you could never contract smallpox. In 1796, a British doctor, Edward Jenner, noticed that milkmaids did not get smallpox. He proved that if he infected someone from the scab of a cowpox sore, that person would never get smallpox. It was the beginning of vaccination. Since that time, medical researchers have developed vaccinations to prevent many diseases, including rabies, diphtheria, tetanus, yellow fever, polio, measles, and mumps.

We are thankful for more than just the milk that cows give us. You could say that because of this docile animal, we have all been spared from catching many terrible diseases.

FACTOIDS

Fortunately, most people recover from chickenpox with no aftereffects. However, it's another story with smallpox. Here are a few interesting facts about this deadly disease:

The Spanish explorer Hernando Cortés brought smallpox to the Americas in 1520 that killed over 3.5 million Aztecs in just two years. It also devastated many Native American tribes.

The Egyptian Pharaoh Ramses V died of smallpox. Five European kings also died of it in the 18th century.

The last natural outbreak of smallpox occurred in Somalia in 1977. The disease has been completely eradicated from the world since that time. The only smallpox virus in the world was stored in two carefully guarded medical freezers, one in Moscow, Russia, and the other in Atlanta, Georgia.

There are other "poxes" named after chickens and other animals such as fowlpox, sheeppox, and swinepox.

DID YOU KNOW?

There is another "pox" that few people have heard of. It is monkeypox, a disease transmitted to humans by monkeys and other primates in Central and West Africa. Monkeypox and smallpox produce almost identical symptoms, including the typical pock marks. Although scientists have known about it since 1958, it was not of great concern. It occurred only in Africa, was typically not fatal, and was not easily transmitted from person to person. There were only 37 reported cases of the disease between 1981 and 1986.

All that changed in 1996 when an outbreak of monkeypox occurred in the Democratic Republic of Congo. Over 500 people caught the disease and 6 of them died. Worse than that, it was discovered that it was being passed from person to person.

Scientists know that if people are vaccinated against smallpox, they will not get monkeypox. However, this presents a major problem. You need smallpox virus to make a vaccine and the only smallpox virus in the world was in two medical freezers.

In 1996, the World Health Organization decided to destroy the last remaining smallpox virus in the research centers in the United States and Russia to totally eliminate smallpox in the world and prevent it from ever occurring again. They decided to keep a half a million doses of smallpox vaccine as well as the smallpox vaccine seed virus in case there is ever a need to produce vaccine in the future. This is stored in the Netherlands.

Let's hope that nothing ever happens to the stored vaccine. If it should accidentally be destroyed, stay away from monkeys.

More questions? Try these Web sites.

Hash House Lingo

http://www.uta.fi/FAST/US8/SPEC/hashhous.html

If you scroll to the bottom of the page you'll find about 15 hash house lingo terms and their meanings. Scrolling down further will give you a longer list of terms and definitions.

The following site will explain many of the terms used in today's restaurants including Gueridon service, table d'hôte, chef de cuisine, and chef de rang.

http://www2.ebham.ac.uk/ca/food%20service/gloss1.htm

Word Oddities and Trivia

http://members.aol.com/gulfhigh2/words.html

A good collection of word trivia, including palindromes (a word spelled the same backward and forward, such as "radar"), names of people that became words, and so on.

The Word Detective

http://www.word-detective.com/

Scroll down the page and click on "Here" under "Looking for back issues?" You'll see hundreds of words and phrases such as balderdash, gandy dancer, gizmo, pig in a poke, honkytonk, scapegoat, the full monty, spitting image, and yahoo. Just click on any word or phrase to find out how the phrase originated and what it means.

City Slang

http://www.slanguage.com/

This site lists local slang expressions used in various cities around the world. For example, you can find out what "gone richter," "orange crush," "the five," and "the boo" mean in Los Angeles, and what "brownie," "egg cream," and "oner niner" mean in New York. The site covers most major U.S. cities plus 18 foreign cities.

The Quotations Page

http://www.starlingtech.com/quotes/

A nice page for lovers of quotations. It includes quotes of the day, motivational quotes, and quotation links. It also lets you search for a particular quotation.

10

Entertainment

What is the origin of the term "Oscar" for the Academy Awards? (It's easier to say than Algernon.)

The official name of the statuette given out at the Academy of Motion Picture Arts and Sciences awards is the Academy Award of Merit. The origin of the nickname Oscar is still not known for sure, but there are a number of stories as to how it came about.

One version is that famed actress Bette Davis told someone that the buttocks on the statuette looked like the buttocks of her husband, Oscar.

The most popular story is that Margaret Herrick, a librarian and later executive director of the Academy of Motion Picture Arts and Sciences, said that the statuette looked like her uncle Oscar. The Academy staff began calling the award Oscar.

After the sixth Academy Awards presentation in 1934, a Hollywood reporter, who had heard the Academy staff using the nickname, mentioned the name "Oscar" when reporting on Katharine Hepburn's award for best actress. It wasn't until five years later that the Academy started using the nickname officially.

The Oscar statuette was designed by Cedric Gibbons who had George Stanley sculpt it. It is the figure of a knight, holding a sword and standing on a reel of film. It was originally solid bronze but after a few years the statuettes were made of an alloy called britannia, which gave the statuettes a smooth finish. Because of a shortage of metals during World War II, the Oscar was made of plaster during that time. When the war was over, anyone who had a plaster Oscar was allowed to redeem it for a gold-plated one.

Today's statuette is 13½ inches high and weighs 8½ pounds. The base is metal, although from 1928 to 1945 the statuette stood on a base of Belgian black marble.

FACTOIDS

The person who has won the most Oscars is not known for being an actor but was someone who has brought joy to children of all ages, Walt Disney. During his lifetime, Walt Disney won 26 Oscars and 6 special Academy Awards. He was also only one of two presenters who opened the envelope to find out that he had won the Oscar (the other was composer Irving Berlin).

After his death, James Dean was nominated twice for best actor, once for *East of Eden* in 1955 and once for *Giant* in 1956. He did not win either time.

In 1941 Orson Welles was nominated for best producer,

director, actor, and screenwriter. He only won best screenwriter, along with his collaborator, Herman J. Mankiewicz.

The Oscar is only one of the awards given out by the Academy. Awards for special achievements may be a scroll, a medal, or any other design. In 1937 a wooden Oscar statuette with a movable jaw was presented to ventriloquist Edgar Bergen for his creation of Charlie McCarthy. Walt Disney received an Oscar and seven miniature statuettes in 1938 when he was honored for his film *Snow White and the Seven Dwarfs*.

DID YOU KNOW?

If you've ever watched the Academy Awards on television, you may have seen someone presenting the Irving G. Thalberg Memorial Award to a "creative producer who has been responsible for a consistently high quality of motion picture production."

Have you ever wondered who on earth Irving G. Thalberg was?

Thalberg, nicknamed "the boy wonder," was one of the most creative producers in Hollywood. However, since he died young, most people outside of the film industry know very little about him.

When he was only 21, Thalberg became head of production at Universal Studios. Four years later he was vice-president and head of production for MGM, which became Hollywood's most prestigious studio. Thalberg helped grind out a series of critical and financial movie successes such as *The Big House, Trader Horn, Grand Hotel, The Merry Widow, Mutiny on the Bounty,* and *The Good Earth.* He also produced *Tarzan the Ape Man* starring Johnny Weissmuller and Maureen O'Sullivan.

Thalberg was about to create his own studio when he died of pneumonia at age 37.

The award that bears his name has been won by top filmmakers in the 20th century, including Darryl F. Zanuck, Hal Wallis, Walt Disney, Cecil B. DeMille, Ingmar Bergman, Alfred Hitchcock, Steven Spielberg, and George Lucas. Darryl F. Zanuck was the only producer to win the award more than once. He won it three times.

Thalberg was respected not only for his filmmaking, but also because he was considered one of the most polite people in Hollywood. He would listen to anyone's opinion and was so modest that his name did not appear on any of his films except the last one he produced, *The Good Earth.*

As the modest boy wonder once said, "Credit you give yourself is not worth having."

Why don't television sets have a Channel 1? (It's all because of an argument about the air waves.)

In the 1930s, virtually all radio was AM, which took up only a small part of the available radio spectrum. Most companies wanted to use the rest of the spectrum for FM.

However, David Sarnoff, president of RCA, had a different idea. He wanted the Federal Communications Commission (FCC) to allocate part of the spectrum to 12 television channels, which would provide three networks to every part of the country.

The FCC was reluctant, because each television station would eat up to 30 times as much spectrum as a single FM channel. The FCC chairman thought the spectrum should be for radio, which the majority of Americans enjoyed, rather than for television, which only the very rich could afford at the time.

Sarnoff tried to force the FCC's speedy approval by demonstrating television at the 1939 World's Fair, hoping it would see the value of allocating room for 12 television channels. Rather than win over the FCC, Sarnoff's brash act angered the chairman, who promptly assigned RCA's proposed television Channel 1 to FM radio.

Today Channel 1 is devoted to FM mobile services such as two-way radios for taxicabs. The other channels were never renumbered, and the original 12 television channels are still numbered 2 through 13. Today there are 68 broadcast television channels, numbered 2 through 69, but still no Channel 1.

FACTOIDS

Prior to 1920, radio was used mainly for maritime, military, and commercial uses. No one considered using it for entertainment. In 1915, David Sarnoff proposed his idea for a "radio music box" to his superiors at the Marconi Wireless Telegraph Company of America. They turned it down. In 1920, he presented the idea to the newly formed Radio Corporation of America (RCA), which agreed to provide the money. Commercial radio was born.

The first licensed radio station in the United States was KDKA in Pittsburgh, Pennsylvania.

Scotland's John Logie Baird gave the first public demonstration of television in 1926, using a scanning system patented in 1884. Baird also adopted the word *television,* coined by Constantin Perskyi in 1900.

David Sarnoff joined with General Electric and Westinghouse to create the first national radio network, the National Broadcasting Company (NBC). It was the first company in the United States created solely to operate a network of radio stations.

There are more television sets in China than anywhere in the world. Over 227 million households have a television set.

DID YOU KNOW?

There are so many inventions that led to television as we know it that it's extremely difficult to answer the question, "Who invented television?"

John Logie Baird was the first to transmit a picture by wireless in 1923, and gave his first public demonstration in 1926. Charles Francis Jenkins, who invented the movie projector, broadcast a 10-minute film in 1925, just two months after Baird's demonstration. Jenkins called his invention "radiovision."

If anyone deserves the title of "television's inventor," it is probably Philo T. Farnsworth. In 1957 he appeared on the popular television show *I've Got a Secret.* His secret was simple. "I invented electronic television—when I was fourteen years old."

With some financial backing, Farnsworth started building his machine. When the device was finished in 1927, he sent his first televised image, a dollar sign, to impress investors. He was just 21 years old.

His innovative design divided the picture into lines of light and shade, and a fast-moving electromagnetic field scanned the picture line by line. That same year he applied for a patent, but it took another 14 years before commercial broadcasting began. However, Farnsworth believed he would eventually make a fortune with his invention.

Unfortunately, World War II broke out in 1941, and in April 1942, all television broadcasting was banned by the government. By the time commercial broadcasting resumed after the war, Farnsworth's patent had run out, and television was in the public domain. Not only did he fail to amass a fortune but his name was not even mentioned as television's inventor. Disheartened, he spent the last 20 years of his life studying nuclear fusion, which he thought could solve the world's energy problems.

Although he did not become rich, Farnsworth contributed a

great deal to society. He held more than 300 patents. He also invented the first cold cathode ray tube and the first electron microscope. He used radio waves to find direction, later known as radar, and developed a black light for seeing at night, used in World War II.

Recognition of his achievements did not come until long after his death. There is now a statue of him in Washington, D.C., and in 1983 the United States Postal Service issued a stamp bearing his portrait.

The next time you turn your television set on, pause a moment to thank a 14-year-old boy who dreamed the impossible but made it happen.

What are the job definitions for all the credits you see at the end of a movie? (Is a gaffer the best boy or not?)

When a movie is over, if you stay to watch all of the credits at the end you'll be there for quite some time. Rather than describe each of these credits, it's better to cover those that most people wonder about.

Gaffer: the head electrician in charge of all lighting personnel. In the early days of film, producers had to rely on natural light. Stages had canvas roofs that could be opened and closed to allow varying degrees of sunlight to fall on the sound stage. Gaffing hooks, traditionally used for landing large fish, were used to move the canvas back and forth. The person responsible for setting the proper amount of light on the stage became known as the gaffer.

Best boy: there are two best boys, one for lighting and one for the grips.The grip is a crew member who works with the camera and electric department to set up and move equipment such as cranes and dollies. One best boy is second in command to the gaffer and the other best boy is second in command to the key grip.

Key grip: the person in charge of everyone who moves anything (grips). Grips move scenery and cameras, set up and take down scaffolding, etc. In live theater they are called stage hands.

Foley artist: the person who creates sounds that cannot be recorded during the filming. Sounds that are later added to the film might be footsteps, creaking doors, thunder, or breaking glass. In radio, they were called sound effects men.

Property master: this person, who is in charge of the prop department, is responsible for obtaining any object that an actor will come in contact with during the film.

Anything that an actor can move is a prop, whether it is a plate, a weapon, or a broom. If the object is never moved, such as a picture hanging on the wall, it is a decoration. For instance, a lamp on a table is a decoration. If an actor is supposed to pick it up and throw it, then it's a prop.

FACTOIDS

The second assistant cameraman is a fancy title for the person who holds the clapboard. The film's title, director, take number, and other information is written on the board. In earlier days, it was written in chalk, but today most clapboards are electronic. The top of the board is slapped shut when filming starts. The sound of the clapboard is later used to synchronize the sound track, while the image is used to synchronize the film track.

Color film was initially viewed as a ploy to attract audiences into the theater. It was often loud, too bright, and unrealistic. Its only purpose was the novelty of being different from black-and-white film. Eventually color became more natural in appearance and was used artistically.

The director of photography does not run the camera, he simply supervises the camera work. He is often called the cinematographer. The person operating the camera is called the "second cameraman."

The credit "craft services" has nothing to do with building models or any other type of craft. This is the group that provides beverages and snacks to the cast and crew throughout the day.

DID YOU KNOW?

There are always people who predict the end of the movie industry. When television first became popular, many people said that it would be the death of film. Yet given more freedom and with newer technology, movies became better than ever.

Today many filmmakers are predicting that film will be obsolete in just a few years and will be replaced by digital media. Some people think that a link of computers, digital video disks, satellites, and television will eventually sound the death knell for movies.

How accurate are these predictions? As Sam Goldwyn once said, "Never make forecasts, especially about the future." In 1922, Thomas Edison said, "I believe that the motion picture will revolutionize our educational system and in a few years will supplant the use of textbooks." Over 25 years ago, some people predicted that the only way movies could compete with television was to be three-dimensional.

It seems that prophets and futurists have been trying to kill off the movie industry for years. However, there's a very good chance that movies are going to be around for a long time to come.

Who said, "Badges? We ain't got no badges. We don't need no badges. I don't have to show you any stinking badges!"

This quote is from the film *The Treasure of the Sierra Madre.* In the film, a band of Mexican bandits approaches Humphrey Bogart and his companions (Walter Huston and Tim Holt) claiming to be *federales,* or Mexican police. When Bogart asks to see their badges, the head of the band says:

"Badges? We ain't got no badges. We don't need no badges. I don't have to show you any stinking badges!"

This quote has been satirized in a number of films. Perhaps the most well known is Mel Brooks's *Blazing Saddles.*

In the film, set in Mexico in the 1920s, Bogart, Tim Holt, and Walter Huston are three impoverished Americans who happen to meet in Tampico, Mexico. They decide to prospect for gold in the Sierra Madre. Houston warns them that greed for gold can change even an honest man.

They find gold and start mining it. As they get more and more gold, they gradually begin to change. Once friends, they now mistrust one another. Bogart initially just wants his "fair share," but in the ensuing weeks this reasonable desire becomes an obsession to get all of the gold. It is a transformation brought about by greed and paranoid fear.

Perhaps the most memorable scene is near the end of the film. Mexican bandits steal the mules loaded with bags of gold dust. The bandits slice open the bags with a knife, reach into them, and think the bags are filled with nothing but sand. They ride off in disgust, as the gold dust trickles to the ground from the open bags. A wind kicks up and the gold dust is blown back into the mountains from where it came.

FACTOIDS

The Treasure of the Sierra Madre was awarded an Oscar for best picture in 1949 and a Golden Globe for best picture of 1948.

John Huston not only had a cameo spot in the film but he also wrote and directed it. He won two Oscars: best screenplay and best director. His father, Walter Huston, won an Oscar for best supporting actor.

The film regularly appears on various "Greatest 100 films of all time" lists.

A small Mexican boy selling lottery tickets in the first part of the film was played by Robert Blake, who later played Little Beaver in the Red Ryder shows and as an adult became a star of both television (*Baretta*) and film.

DID YOU KNOW?

In the film *The Treasure of the Sierra Madre,* three men are hunting day and night for a treasure that lies buried on the other side of the next hill. They sought gold, but the true treasure of the Sierra Madre is the native culture, the Huichol. These Indians have lived in the isolated Sierra Madre for centuries and offer the world the wisdom of the ancients.

The Huichol have been separated from the modern world by the rugged Sierra Madre. They call themselves "the healers" and perform rituals that they believe heal the Earth and keep all of nature in balance. A key ceremony is the sacrifice of a white-tailed deer, the blood of which nourishes the earth.

Although the Huichol were isolated from civilization for centuries, the modern world invaded their realm in the 1970s. Because the Mexican government wanted to integrate all cultures into the mainstream of society, schools, clinics, and government agencies intruded on the land of the Huichol. Ranchers wanted to graze their cattle on the grassy plateaus of the Huichol land. Missionaries wanted to convert the "pagans." Tourists and government officials started arriving. In spite of all this, the Huichol clung to their ancient ways.

Major cities drained water from the Sierra Madre. Soon forests vanished and game died. Poverty and illness plagued the Huichol. As they heard of environmental pollution in other parts of the country, they blamed themselves. There were no more white-tailed deer in the forests and they could no longer perform the ancient ceremony. That was why the earth was sick. They knew they had to find a way to heal the earth.

The Huichol made an incredible journey. They left their mountain homeland and made a 600-mile pilgrimage to the heart of Mexico City, the largest and most polluted city in the world. They were hoping to obtain some white-tailed deer from the zoo so they could perform their sacred ceremony. They thought it was their duty to save the earth for the benefit of humans everywhere.

The devoted Huichol got their white-tailed deer. In 1988 they were awarded the National Ecology Prize of Mexico. But will their selfless acts and ancient ceremony save the earth from environmental destruction?

Only time will tell.

Where can you find the original words for Taps? (Fare thee well, day has gone.)

Bugle calls were designed to give signals to soldiers. They originated because the sound could be heard over the din of battle; they were not intended to be sung. In modern times, however, people have written words for well-known bugle calls.

It is believed that Taps was part of an earlier French bugle call, the Tattoo. The word *tattoo* originated in the middle of the 17th century. Soldiers used to spend their evenings at taverns near the army post. When it was time to return to the base, military police would sound a signal to tell the tavern owners to shut the taps on their kegs and the soldiers to return to the post. *Tattoo* started out as the Dutch word *taptoe* (*tap,* meaning "faucet," and *toe,* meaning "to shut"). When shifting from Dutch to English, the word *toe* become "too," and then "taptoo." The name Taps comes from "taptoo." Further corruption of "taptoo" resulted in the word *tattoo,* which is still used today.

Anyone who has ever heard Taps knows it is an eloquent and haunting tune, yet its origin is still argued today. The most widely accepted explanation is that it was created by a Civil War general, Daniel Adams Butterfield.

At that time the Union Army used the French final bugle call Lights Out to signal the end of the day. Butterfield thought it was much too formal. He recalled Tattoo, instructed his aide to write it down, and had the brigade bugler play it. Butterfield made some notes longer and some shorter but did not deviate from the original melody. He then had the bugler sound Taps at the end of every day.

Butterfield's rendition of Taps is the one we still hear today being played at the end of the day, at funerals, and at memorial services. With only 24 notes, this poignant and eloquent bugle call still stirs the emotions of all who hear it.

FACTOIDS

When England's King George III introduced bugle calls, he had a special call written for waking the troops. The call was named *Reveille,* from the French for "to wake."

The soldiers in the Roman legions were awakened by horns playing Diana's Hymn. Today the French term for Reveille is *La Diana.*

There are 25 bugle calls used at a typical Army installation today. Of these, 20 are almost always used and 5 are optional. The more common calls are: First Call, Reveille, Assembly, Mess Call, To the Color, Retreat, Tattoo, and Taps.

The bugler's First Call is the same melody commonly heard today at the start of horse races.

Taps was first played at a military burial because the commanding officer thought that the traditional three-volley salute would provoke the enemy encamped nearby.

DID YOU KNOW?

When recalling the history of Taps, it's easy to conjure up a mental picture of the famous 7th Cavalry coming to the rescue of besieged settlers, the blaring bugler at the front of the charge.

If Taps ever moved anyone, the soldiers of the 7th Cavalry must have been choked with emotion when they heard the bugle play for over 200 of their dead comrades who had been annihilated by the Sioux in the Battle of the Little Big Horn, commonly known as "Custer's Last Stand."

Only one member of the army survived the battle, but it wasn't a soldier. It was a horse, named Comanche, who had been ridden by one of Custer's officers. Comanche's body was pierced by seven arrows, but he was given medical attention and nursed back to health until he fully recovered. Given the freedom of the fort grounds, Comanche was always saddled for official occasions, but the commanding officer ordered that he never be ridden again. When Comanche died, newspapers throughout the country published his obituary.

After his death, Comanche was stuffed and put on display in the Museum of Natural History at the University of Kansas. He's still there today.

Some anonymous author composed five verses for Taps.

Although Comanche couldn't read, he probably would have fully understood the last comforting verse:

> Thanks and praise, for our days,
> 'Neath the sun, 'neath the stars,
> 'Neath the sky.
> As we go, this we know,
> God is nigh.

What is the name of the dog in the Maytag commercials? (I don't give a fig for this one.)

The canine companion of the "Lonely Maytag Repairman" or "Ol' Lonely" is a basset hound named Newton. The word *basset* is from the French for "low," and basset hounds are definitely that.

The first lonely Maytag repairman was played by veteran character actor Jesse White, who died of a heart attack in 1997 at age 78. He had been a regular on four television series, including *Private Secretary* and *The Danny Thomas Show,* and was the voice on three other television series. He also appeared in many other television shows and movies.

Since 1989, the lonely Maytag repairman has been played by Gordon Jump, probably best known for his role as the general manager of *WKRP in Cincinnati.* He has been in 4 television series and 25 movies, including *Conquest of the Planet of the Apes.* He has made guest appearances in over 40 television shows, including *Seinfeld, Caroline in the City, Love Boat, The Brady Bunch,* and *Get Smart,* to name just a few.

In Canada's French-speaking Quebec, the lonely Maytag repairman is played by Paul Berval.

FACTOIDS

Although a basset hound may look strange, there is a reason for its appearance. It is trained as a hunter, and its long ears stir up scents on the ground that are detected by its large nose. The folds of skin under its chin help retain the scent. It is a sturdy dog because of its large feet and heavy bones. Short legs let the hunter keep up with him when chasing prey.

Basset hounds are very lovable and are often called "armchair clowns" because of the funny positions they take when sleeping.

Very few dog breeds love people as much as basset hounds do.

Because a basset hound has very short legs and two-thirds of its weight is in the front part of its body, it can swim only very short distances with great difficulty.

DID YOU KNOW?

Perhaps you don't often think about what a wonderful invention the washing machine is. Before its invention, people used sand as an abrasive to free the dirt from clothes and then washed them by first pounding them on rocks and then washing the loosened dirt away in streams. The scrub board was not invented until 1797.

It took another 75 years before the first washing machine appeared. An Indiana merchant, William Blackstone, built it as a birthday present for his wife. He took a large wooden tub and put in a flat piece of wood with six small pegs. When a handle was turned, the pegs snagged the clothes and swirled them about in hot soapy water.

Blackstone's idea caught on, and competitors started copying his machine and adding improvements such as metal tubs and a wringer. Just one year after Blackstone built his machine, there were over 2,000 patents for various washing devices.

In the first part of the 20th century, new methods of powering the washers were tried. At first, gasoline engines were used; later, electric motors became the power source. An enterprising California gold miner and carpenter built a machine that could wash 12 shirts at a time. The power to run it was supplied by 10 donkeys.

Inventors of washing machines tried different methods to get the dirt out of clothes. Some machines had rollers that squeezed out dirt, some had mechanisms that "stomped" the clothes as if they were pounded on a rock, and some slammed the clothes against the sides of the tub to free the dirt. All of these devices dragged clothes through the water. It wasn't until 1922 that the Maytag Company created a device that did just the opposite: it forced water through the clothes. The device was called an agitator.

Of all of the methods tried, only two still exist today, the cylinder system that tumbles clothes inside a moving tub and the agitator.

Today's washing machines have multiple cycles and features that let you wash clothes as efficiently and quickly as possible.

The next time you wash your clothes, rather than getting upset about how much work it is, just remember one thing. It's a lot better than pounding them on a rock.

More questions? Try these Web sites.

Boy Scout Song Book
http://members.iinet.net.au/noneilg/scouts/songs

Contains collections of various songs that can be sung by scouts or at campfires. This Web site has links to a great variety of songs, not only from the United States but also from Australia and South Africa. You can also download complete songbooks if you'd like.

Internet Movie Database

http://us.imdb.com/

This site contains the names of the top movies of the week, news features, upcoming movies, and a great deal of other information about movies. You can type in your city or zip code and find out local show times.

You can search the huge database for just about any film that's been made. It will give you a short plot description as well as a complete cast list.

Where Are They Now?

http://people.aol.com/people/where_now/

If you want to know what yesterday's famous movie, television, and music stars are doing today, this is the site for you. Just click on "Movie Stars," "TV Stars," or "Music Stars" and then click on the celebrity you want information about. For example, you'll learn that Gary Burgoff (Radar O'Reilly in M*A*S*H) now earns a living selling his paintings of animals which can bring as much as $25,000 each.

Seeing Stars

http://www.seeing-stars.com/

If you plan to make a trip to Hollywood, check out this site before you go and you'll stand a much better chance of running into a movie star. It lists where they shop, where they eat, where they play, where they go to school, and where they go to church.

The site also lists Hollywood landmarks, famous streets, and how to find the locations where specific movies have been filmed.

THE NETWORK AND CABLE TV GUIDE
http://www.geocities.com/TelevisionCity/9348/tv_guide.htm

A wonderful site for the television enthusiast. It has detailed information on just about every show every televised. Scroll down to select a program by name or by star. The site also lets you hear music from some of the more popular shows. The site includes prime time schedules from 1970 to 1994, trivia, and links to other sites.

11

Science

Why are there C and D batteries, but no A and B batteries? (Flashlights and model trains have a lot more in common than electricity.)

The letters assigned to battery types were selected by the American National Standards Institute (ANSI) in the 1920s.

Although there were A and B batteries at that time, the only sizes that caught on in a big way commercially were AAA, AA, C, and D.

Sizes available today range from AAAA to G, and there are also J and N batteries.

If you were to take apart a large 6-volt lantern battery, you'd find it actually consists of four F batteries.

A battery is nothing more than a container for a substance that produces electrons. There are both wet and dry batteries. In both cases the battery contains some type of chemical that acts as a conductor of electricity. Wet batteries contain a liquid chemical, while dry batteries contain a moist paste.

In concept, a battery is a rather simple device consisting of two electrodes that are inserted into a container holding chemicals that produce electrons. One electrode, or terminal, is positive and the other is negative. As the chemicals produce electrons, they accumulate on the negative terminal. It is said that opposites attract, and electrons are no different. They are attracted to anything positive. If something is connected between the terminals, such as a small electric motor, electrons on the negative terminal are attracted to the positive terminal and flow to it as quickly as possible. This flow of electrons is called electricity, which passes through the motor and starts it running.

Most batteries we are familiar with are "dry cell" batteries, used in flashlights, wristwatches, calculators, and other similar products.

The dry cell battery consists of a metal container filled with a moist paste. A metal electrode or graphite rod is inserted into the center of the paste and the container is capped. The metal case itself is the negative terminal and the rod is the positive terminal. When you look at a battery and see the "bump" on top, it's actually touching the top of the rod. When something such as a miniature light bulb is connected between the case (the metal is usually exposed only at the bottom of the battery) and the "bump," electricity will flow through the bulb and light it.

When the flashlight was first invented, the cover of the

Eveready Battery Company catalog displayed the biblical quote, "Let there be light."

FACTOIDS

The Eveready Battery Company, now called Energizer Holdings, Inc., produces more than 6 billion batteries every year.

Batteries will not last longer if they are stored in a refrigerator or freezer. In fact, they probably won't last as long because the cold and condensation will damage them.

Of all the batteries sold, one out of every five is used in flashlights.

Batteries have a long history. Over 200 years ago, in 1799, Alessandro Volta invented the battery (guess where we got the word "volt"). The flashlight was invented 100 years later, in 1899.

DID YOU KNOW?

Joshua Lionel Cowen founded the American Eveready Company, yet some people said he was a very foolish man.

Cowen wanted to revolutionize the photography of his time. He invented a fuse to set off the photographic flash powder, but it didn't work very well, at least not for photography. However, the U.S. Navy bought 24,000 of the fuses to set off underwater explosives.

He next tried creating lighted flowerpots, but had so much difficulty perfecting the design that he got bored and gave away both the project and his company to a salesman named Conrad Hubert.

Hubert wasn't interested in flowerpots at all, but he liked the idea of a metal tube with a light bulb and a dry cell battery. He modified the basic idea and came up with the world's first flashlight. The idea that Cowan gave away made Hubert a multi-millionaire. That's why some people called Cowan a foolish man. But that's not the end of the story.

Cowan was still fascinated with electrical devices and created a window display to draw customers into a store. It was a battery-operated toy train that sped around a circular track. There was a problem, though. Onlookers came into the store to buy not the sale merchandise but the display. Many thought the train would look wonderful under the Christmas tree. So Cowan started a company to make model trains and ended up making a fortune. He used his middle name for the company name.

You guessed it. It was the Lionel Train Company, perhaps the most famous model train company that ever existed.

If you could hollow out a sphere in the exact center of the earth large enough to hold a person, would that person feel gravity? (Hot and heavy.)

We know that in reality this would not be possible, because the temperature at the center of the earth is around 9,000°F, hotter than the surface of the sun, and the pressure is more than 3.5 million times that of our atmosphere. So it would just be a question as to whether you burned up or were crushed first.

However, if you could be put in the center of the earth, the gravity from the earth's mass would pull you in every direction at the same time. In essence, there would be no gravity. You would not be able to distinguish among up, down, or sideways. You would have a sense of weightlessness, similar to that experienced by astronauts in outer space.

Suppose you dug a tunnel down to the center core and then dug another, parallel tunnel back up to the surface. If you were to use the first tunnel to get to the center of the earth, you would have an exciting ride.

As you started falling down the tunnel, you would go faster and faster until you reached the center of the earth. Because of your momentum you'd then start shooting up the second tunnel. However, as you got farther from the center, gravity would start slowing you down. You'd almost reach the surface before falling back again, going faster until you reached the center, when you'd start zooming up the first tunnel. But this time gravity would slow you down even more, so you'd be farther from the surface this time. You'd do this over and over and over again, traveling less each time until you finally came to a stop, suspended at the center of the earth.

It's a wild ride that is definitely not recommended.

FACTOIDS

If the earth's diameter were 10 percent larger or smaller, all life would either burn up or be frozen.

The earth weighs over 6.5 sextillion tons. That's 6 with 21 zeroes after it.

Not all the dust you see comes from the earth. Some scientists estimate that over 30,000 tons of cosmic dust are dumped on our planet every year.

The earth's axis is tilted about 23° from the perpendicular. If the earth's axis were perpendicular, we would have no seasons and it would be impossible to grow food.

DID YOU KNOW?

In Jules Verne's book *Journey to the Center of the Earth,* a professor, his nephew, and an Icelandic guide enter a dormant volcano and travel to the earth's center. It's a fascinating idea, but could such a journey be possible? Believers in the "hollow earth theory" think it could be.

It all started in the 17th century when the famed scientist and discoverer of Halley's comet, Edmund Halley, theorized that the earth was composed of concentric spheres, one inside another, and that each of these spheres might support life. Although modern science has pretty much proven how the earth is constructed, the hollow earth theory did not die.

A few hundred years after Halley's theory, other people claimed that they could prove the earth was hollow. There were large openings at the poles, they said, that would allow people to gain access to the center of the earth.

In 1947, Admiral Richard Byrd flew over the North Pole. An alleged "secret diary" of Byrd's claims that he encountered green rolling hills and a living mammoth-like creature. Even Hitler's top advisers believed that the earth was hollow and mounted an expedition to see if they could enter the hollow earth as a means of escape should they lose the war.

There are hundreds, if not thousands, of documents that discuss mysterious underground caverns and tunnels crisscrossing the earth, as well as ancient stories of beings living inside the earth.

Although scientists today scoff at these beliefs, there are still many people who believe in the hollow earth theory. Although the proponents of this theory have many fascinating arguments, they aren't convincing. After all, there are also people who still believe the earth is flat.

Has a tenth planet been discovered recently in our solar system? (What about a death star?)

Scientists often use the term "minor planets" when referring to asteroids. A few years ago an amateur astronomer in England discovered a new asteroid in the asteroid belt. There was nothing unusual about this because many new asteroids are discovered every year.

A newspaper picked up the story, but instead of reporting that a "minor planet" had been found, someone left out the "minor," so that the newspaper story said a new planet had been found. It wasn't long before newspapers, radio stations, and television stations around the world were reporting the discovery of a new planet. Fortunately, the truth eventually became known.

Some astronomers believe that a tenth planet does exist; other scientists do not. Although we have not discovered any new planets in our solar system lately, scientists have discovered 14 new planets in other solar systems.

The search for a tenth planet goes on. Many scientists believe that there is a new planet out there. Some theorize it's about half a light year away (3 trillion miles) and could be up to 10 times as large as Jupiter. If you were to draw a diagram of the solar system so that the Earth is just one inch from the sun, Pluto would be a yard from the sun, and this theoretical tenth planet would be half a mile away, too dim to be seen by any telescope we have today.

No one has bothered to name this mythical planet. Vulcan and Planet X have already been used in the past to name theoretical planets. If the planet is found, whatever name it is given will surely go down in history.

FACTOIDS

When the planet Pluto was first discovered, some scientists suggested names such as Atlas, Zymal, Perseus, Vulcan, Tantalus, and Cronus. Newspaper reporters suggested Osiris, Bacchus, Apollo, and Erebus. Other people suggested Minerva, Constance, and Zeus. A few months after its discovery, the planet was officially named Pluto, a name suggested by Venetia Burney, an 11-year-old schoolgirl in Oxford, England.

In the 1600s, it was believed that Venus had a satellite. It was seen over 33 times by 15 different astronomers for almost two centuries. It has never been seen again since 1768, and scientists today are convinced it doesn't exist. However, the observers of the satellite were astronomers who had reputations for accuracy and reliability. It's difficult to believe that so many reputable scientists over a period of almost two centuries could all make exactly the same error. Even today, it is still a mystery. Did it exist? And if it did, what happened to it?

Scientists have discovered over 120 meteor craters on the earth, ranging in size from a few hundred feet in diameter to 100 miles in diameter. The original meteorites are still at the bottom of some craters. However, scientists believe that most meteor craters haven't been discovered yet.

A spectacular meteor train was seen in Canada in 1913. A fiery red body was first seen over Saskatchewan and observed until it passed over Bermuda. As soon as it had passed, a number of smaller bodies with tails followed in groups of twos or threes. There was a procession of 20 major bodies along with groups of 40 or more smaller bodies over a 2,500-mile path. None of the meteors hit the earth and no one knows what happened to them after they passed Bermuda.

DID YOU KNOW?

Is it possible for a "death star" to exist? Some scientists believe one does. They have theorized that a faint or dark star moves in an elliptical orbit around our sun. This dark star, or "death companion" passes through a cloud of comets at a very long distance from the sun, stirs them up, and causes earth to collide with these comets tens of thousands of years later. This is theorized to happen around every 30 million years or so. The scientists who proposed this theory in 1985 named the sun's companion star "Nemesis," which means a rival or something that inflicts vengeance.

The earth's geological record indicates that most life has been annihilated every 30 million years, as when the dinosaurs became extinct 65 million years ago.

The idea of a virtually invisible sun using comets to attack the earth sounds like a science-fiction story and most scientists do not accept the theory. However, they admit that it is possible.

If the theory of this death star eventually proves to be true, that means we are due for another mass extinction of life on earth. Should we worry? Probably not. It isn't scheduled to happen until about 15 million years from now.

How are seedless watermelons produced? (No more world's record for seed spitting.)

Watermelon seeds can be planted to grow more watermelons. However, if you have a watermelon with no seeds, how can you possibly grow any more watermelons? Well, it is possible, thanks to modern science.

First of all, consider what happens when a donkey and horse are mated. The offspring, a mule, is sterile because it is genetically different from either of its parents. To get another mule, you have to mate a donkey and a horse again.

It's the same with watermelons. If two watermelons are genetically different, their offspring will look like normal watermelons, but they will be sterile and won't produce seeds.

Normal watermelon plants have two sets of chromosomes. The first step in producing a seedless watermelon is to extract a compound, called colchicine, from autumn crocus plants. The compound is added to the watermelon plant, doubling the chromosomes so that the plant now has four sets.

The next step is to cross a plant with four sets of chromosomes (a female plant) with a normal plant that has two sets of chromosomes (a male plant). Each plant then gives half its chromosomes to the offspring. The male produces pollen with one set of chromosomes, and the female uses two sets to produce an egg. All the offspring will each have three sets of chromosomes.

The rind and flesh of a watermelon grow from the plant's nonreproductive cells. So the offspring grow into normal plants and produce seemingly normal watermelons, although they cannot produce seeds. The plant has only three sets of chromosomes, but it needs four in order to produce seeds.

In the past, farmers had to buy seedless plants from breeders and then transplant them into the fields. In addition, they had to rely on bees, because even seedless plants have to be pollinated. Farmers would plant seeded and seedless plants next to each other so that the bees would pollinate the seedless plants in order to produce seedless watermelons.

Now scientists have figured out a way to produce seeds from seedless plants. These seeds will produce seedless watermelons. However, they are expensive, costing as much as $150 per 1,000 seeds or 15 cents for a single seed.

Many people like the new seedless watermelons. Others

miss being able to spit out the seeds. Hopefully, there will always be watermelons with seeds. If not, there won't be any more watermelon seed–spitting contests.

FACTOIDS

In Florence, Italy, there is a celebration every year on August 10 to commemorate the patron saint of cooks, Saint Lorenzo. During the event, people go on a watermelon-eating binge.

Watermelons were not grown commercially in the United States until the 1930s, although they were eaten in Egypt over 5,000 years ago.

A traditional watermelon can have as many as 1,000 seeds.

During the Civil War, the Confederate Army boiled down watermelons as a source of sugar.

Lee Wheelis of Luling, Texas, holds the world's record for watermelon-seed spitting, with a record 68 feet 9 inches.

DID YOU KNOW?

Seedless watermelons were created because of advances in modern science. But hundreds of agricultural miracles were performed by a man who was born in 1849, Luther Burbank.

Although he never went to college, Burbank was an avid reader and was greatly impressed with Charles Darwin's treatise "The Variation of Animals and Plants Under Domestication." When he was 21, he bought a 17-acre piece of land and started his experiments, which he continued for the next 55 years.

Luther Burbank conducted as many as 3,000 experiments at once. He knew nothing of genetic principles. He experimented by grafting and crossbreeding native and foreign plants to produce new plants. Even today, some of his early inventions are still with us, including the Shasta daisy, the Elberta peach, the Santa Rosa plum, and the Flaming Gold nectarine.

When he was only 22 years old, he created the Russet Burbank potato, now commonly called the Idaho potato. It was exported to Ireland to help the Irish recover from the great potato famine of 1840 to 1860. The Idaho potato has become a staple of agriculture in the United States and has remained unsurpassed to this day.

Luther Burbank developed over 800 strains and varieties of plants during his lifetime. Of the 113 varieties of plums and prunes he developed, 20 are still commercially important today. He also developed 10 commercial varieties of berries and 50 varieties of lilies.

The next time you bite into a delicious peach or plum, there's a good chance you are tasting one of Luther Burbank's discoveries.

Is it true that glass is a liquid? (Liquids don't break.)

People used to think that glass was an extremely thick liquid. They got the idea from glass windows in ancient cathedrals, because the windows were thicker at the bottom than at the top. They believed that glass flowed like a liquid, only extremely slowly, and that's why the glass eventually became thicker at the bottom.

Scientists decided to test this idea and discovered that to see any significant flow in glass, you'd have to wait at least a hundred million, trillion, trillion years. That's a very long time. It turns out that old cathedral windows are thicker at the bottom because of the way they were made. It's also believed that the early glass makers intentionally made windows thicker at the bottom for stability and safety.

Because glass is definitely not a liquid, it must be a solid. No, that's not true either. So if it's neither a liquid or a solid, and it's definitely not a gas, what is it?

Although the atoms in glass stay in one place just like atoms in a solid, they are arranged in the jumbled fashion of a liquid. Therefore, glass is sometimes described as a liquid that has cooled and thickened to the point where it has become rigid. In other words, it's a frozen liquid. However, that's not quite right either because a liquid flows when subjected to stress and unheated glass cannot flow.

The best way to describe glass is that it's a shapeless, or noncrystalline, solid. In a crystalline object, the atoms line up perfectly. However, in glass, the atoms are all mixed up like a bunch of loose balls in a big box.

The main ingredient of glass is silicon dioxide. The most common form of silicon dioxide is sand. To make glass, the sand must be heated until it melts. However, it's impossible to get a fire hot enough, unless it's a meteorite. Therefore, a chemical called a flux is added to help the sand melt at a lower temperature. The Romans discovered that you could lower the temperature for making glass by adding lime and soda to the sand.

If you were simply to melt sand, you would create glass, but it would tend to crumble or break easily. Therefore, a chemical called a stabilizer is added so that the glass will be uniform and retain its basic structure.

The basic ingredients of glass are called formers. In addition to silicon dioxide, the basic mixture might include various chemicals to color the glass. Various combinations of formers, fluxes, and stabilizers are used to create thousands of different types of glass.

Glass has been used for thousands of years to make containers as well as myriad jewelry products. Today it is also used for windows, mirrors, lamps, vases, and thousands of other products.

Wherever you are at the moment, look around and you'll probably see a number of glass products.

FACTOIDS

Because glass takes one million years to decompose, it never wears out and can be recycled almost forever.

Laminated glass is created by bonding sheets of tough plastic between layers of glass. Automobile windshields are made this way. Thicker sheets of glass are laminated to create bulletproof glass.

Milk glass, sometimes called porcelain glass, is an opaque white glass made by adding burned bone or horn to the glass mixture. It was created as an inexpensive substitute for Chinese porcelain.

If lead oxide is added to the mixture when glass is being made, the result is called crystal. Crystal was originally called flint glass, because flint was used instead of lead oxide. Crystal is known for its brilliance and clarity.

DID YOU KNOW?

Almost every child has played with marbles at one time or another. In ancient times, marble games were played with rounded pebbles, nuts, or fruit pits. In modern times, marbles have been made from baked clay, steel, plastic, glass, and other materials.

Most modern marbles are made from glass. To make a handmade marble, a rod of glass is used. The rod usually consists of smaller rods of different colors. The rod is heated and gently twirled until a round blob is formed. The blob is then cut off.

When marbles are made by machine, a stream of molten glass is shot out of a furnace through a small hole. As the stream emerges, a continually revolving scissor cuts off the exact amount needed for a single marble. The blob of molten glass falls onto a belt made of counter-rotating rollers. Because the rollers are inclined, the glob travels down the center of the belt and eventually forms a perfect sphere. By the time it reaches the end of the belt, it has cooled and become a marble.

In recent years the game of marbles has been making a comeback. A number of national tournaments are held in the summer. So get your best "aggie" shooter and "knuckle down" to the line.

In movies, why does a wheel appear to be rotating backward when the wagon or car is moving forward? (It happens to airplane propellers also.)

Most people have seen movies in which a wagon or stagecoach is traveling through the country and the wheels look like they are moving backward even though the stagecoach is moving forward. You won't see this phenomenon in real life, but you will often see it in movies and television. The illusion occurs because a film is actually a series of still photographs, replayed so fast that your eye thinks it is seeing a continuous flow of images.

A single still picture is called a frame. The eye can see approximately 12 frames a second. To fool the eye into believing it's seeing motion, frames must be shown at least twice as fast as the eye can see them. Movie films display 24 frames a second and television films display 30 frames a second.

To understand this illusion, assume that the wheel rotates once a second and you show a still picture once a second. In this case the wheel will appear to stand still. This is because in the interval between each frame a new picture moves into the same position as the previous picture. However, if the frame interval your eyes see doesn't match the camera's frame interval, the wheel will appear to move either forward or backward.

If a frame on the film arrives a little sooner than the camera's frame interval, the wheel appears to move forward. On the other hand, if the frame arrives a little late, the wheel appears to move backward.

The next time you see an automobile in a film, watch closely as it speeds up. Although the car is continually moving forward, the wheels will first appear to be going backward, then they seem to stop, and finally they appear to be going forward.

On the other hand, if you really want to enjoy the movie, keep your eye on the car instead of the wheels.

FACTOIDS

Stop-motion or stop-action photography is also used to fool the eye into seeing motion. A still photograph is made of an object, such as a clay model of a dinosaur. The object is moved slightly and another photograph is taken. This process is repeated thousands of times. When the photographs, or frames, are shown at the speed of a motion picture camera, 24 frames per second, the clay model appears to be moving. Here are a few interesting facts about stop-motion photography.

During the photo session, if a model is accidentally moved, it could mean that the entire scene, consisting of thousands of photos, would have to be redone. To avoid this problem, the camera, set, and models are usually clamped down except when shooting. During the making of *King Kong,* the miniature foliage was made of metal so it couldn't easily be moved by accident.

A major problem with stop-action filming is that there are no "blurs." If you film a man running down the street, there will be a slight blur on each frame. Although not noticed by the audience, the blur helps make the running motion smooth and realistic. In stop-action films, a running creature or person always seems to have jerky movements. This problem has been solved with computer animation, which can be used to blur frames to produce realistic movement.

In one scene of *King Kong,* a live primrose plant was used in a jungle scene. No one noticed that the plant was blooming. When the film was reviewed, a white primrose opened up in time-lapsed fashion. An entire day's work had to be scrapped.

DID YOU KNOW?

Stagecoach wheels may go forward or backward, but one stagecoach turned Hollywood completely around: the 1939 film, *Stagecoach.*

Prior to that time, countless Westerns had been churned out by cheap studios, and no one in Hollywood took Westerns seriously. In 1939, director John Ford proved that a Western could be intelligent, entertaining, and profitable. The film started the era of quality Western films and launched the career of a young actor who had his first starring role in *Stagecoach.* The actor's name was John Wayne.

Stagecoach is considered by many to be one of the top 50 films of all time. In 1939 it was nominated for seven Academy Awards, including best picture, best actor, and best director. Since 1939 is usually considered to be the best year in film history, at least 15 of the films eventually became all-time classics. A few of the films released in 1939 included *Gone With the Wind, The Wizard of Oz, The Hunchback of Notre Dame, Of Mice and Men, Gunga Din, Beau Geste,* and *Mr. Smith Goes to Washington.*

Although *Stagecoach* only won Oscars for best supporting actor and best score, it's well worth the effort to rent the video and relish this classic movie.

More questions? Try these Web sites.

SHOOTING MARBLES

http://www.marbles.net/

If you scroll down and click on "Links," you'll find places to buy both homemade and machine-made marbles, links to sites devoted entirely to marble collecting, places to buy display cases for your marbles, tournament rules, and a list of more marble links if you click on "Friends places." There are even links to sites for kids.

HOW STUFF WORKS

http://www.howstuffworks.com/index.htm

Have you ever wondered how a jet engine, a refrigerator, a thermos, or a cell phone works? If so, you'll find the answer on this site. Just click on any one of the 25 categories to see numerous explanations of how things work.

DISCOVERY CHANNEL

http://www.discovery.com/

This is the official Web site of television's Discovery Channel. Descriptions and video clips of Discovery Channel programs, science news, and fun and games for children are just a few of the many features on this site.

NOVA

http://www.pbs.org/wgbh/nova/

This site features NOVA, a science and adventure program broadcast on public television stations (PBS) throughout the country. In addition to featuring programs already broadcast, it has a list of programs, television schedules, and a station finder to tell you where NOVA is shown in your area.

NIKOLA TESLA

http://home.nycap.rr.com/useless/tesla/tesla.html

Nikola Tesla was a genius who gave us alternating current, the fluorescent bulb, and neon lights. Tesla also demonstrated radio principles 10 years before Marconi, and gave us many other innovations. Although considered a genius by many, he was thought to be crazy by others.

In spite of his scientific contributions, Tesla said he had communicated with Mars, talked about death rays with a range of 250 miles, supposedly caused an earthquake for blocks around his laboratory, and created the largest ever man-made lightning bolt.

Today, Tesla is becoming recognized for the genius he was. This site is a "must read" if you're interested in science. If you want even more information about Tesla and his inventions, go to:

http://inventors.about.com/science/inventors/library/bl/bl1_ 3t.htm. Scroll down and click on "Nikola Tesla."

12

Sports and Games

How did they pick the name Yahtzee for the dice game? (You don't need a yacht to play.)

It all started in the 1950s when a wealthy couple from Canada invented a game they could play aboard their yacht. Whenever the couple entertained guests on their yacht they taught them how to play the "yacht game."

In 1956 the Canadian couple asked Edwin S. Lowe to make some samples of the game to be given as gifts. Lowe had amassed a fortune selling bingo games 30 years before.

Lowe liked the game so much that he offered to buy the rights. The originators of the game sold him the rights in return for 1,000 games they wanted to give to their friends. He eventually changed the name from the "Yacht game" to "Yahtzee."

Initial sales of the game were very slow. It was difficult to try and explain Yahtzee in an advertisement. You almost had to play the game to understand its appeal. That gave Lowe a brilliant idea. He starting having Yahtzee parties, and word of mouth eventually led to substantial sales of the game. Today, more than 50 million Yahtzee games are sold each year.

Many variations have sprung up over the years such as Triple Yahtzee, which is similar to playing three Yahtzee games simultaneously, Painted Yahtzee, which is a colored version of the game, and Battle Yahtzee, which lets you earn a chance to throw your dice at your opponent's combinations (this could easily start a battle).

You don't have to own a yacht to have fun playing the "yacht game."

FACTOIDS

An estimated 100 million people play Yahtzee every year.

One of the earliest board games was shaturanga, which was invented in the 6th century by an Indian philosopher. Many authorities believe it was the predecessor of chess.

The odds of rolling Yahtzee in one roll are 280,000 to 1.

Backgammon was played by Egyptians, who called it the game of 30 squares. The rules they played by are not known today. The ancient Romans called it the game of 12 lines. A variation of the game was played in Asia before A.D. 800.

DID YOU KNOW?

Edwin Lowe not only made Yahtzee popular but was responsible for one of the most popular games in the country today, bingo.

The game of bingo evolved from an Italian lottery game created in 1530, called Le Lotto. By 1700 the game had migrated to France and became an educational tool in Germany in the 1800s. It was used to teach children multiplication tables.

The American form of bingo has been called keno, kino, pokeno, beano, screeno, lucky, and fortune.

In 1929 people were playing beano at a carnival in Georgia. Each player had a piece of cardboard with numbers on it and placed a bean on the number if it was called out. The first player to fill a line on his card would yell "beano!"

Lowe had watched people playing beano and introduced the game to some of his friends. One of his players filled a line and was so excited that he yelled "Bingo!" instead of "Beano." Lowe decided to call the game bingo and competitors paid him one dollar per year for rights to use the name.

A year later Lowe hired a Columbia University mathematics professor to find out if he could increase the quantity of number combinations. The elderly professor created 6,000 bingo cards with nonrepeating groups of numbers. He then went insane!

A Catholic priest asked Lowe to set up a game for him for a church function because by now Lowe was producing bingo cards day and night. Within a few years bingo had spread along the eastern seaboard and started moving west. By 1934 over 10,000 bingo games were played a week and by 1995 players spent a total of $88 million a year on bingo. It is estimated that in a single year players shouted "Bingo!" 1.2 billion times in the United States alone.

Whether bingo is played in a church hall, a lodge hall, or a Las Vegas casino, it looks as if it is here to stay.

Why do golf balls have dimples? (It's a real drag.)

If you were going to design an object with a wonderful aerodynamic shape, you would never design a ball. Because a ball is round, air will flow smoothly around the front half of the ball when it's in flight. However, as the air flows behind the ball it causes turbulence, which causes drag and slows the ball down. A slower ball means a shorter flight.

To be just a bit more scientific, when the air pressure in front of the ball is significantly higher than the air pressure behind the ball, drag occurs. A solution to the problems of an aerodynamic ball would be to somehow increase the pressure behind the ball so it approximates the pressure in front of the ball. Because of the dimples in a golf ball, air flowing around the ball is less turbulent, because each dimple creates a smaller area for the turbulence and there is much less drag.

To put it simply, dimples in a golf ball greatly reduce drag and the ball flies farther. In fact in some studies, a dimpled ball flew four times as far as a smooth ball.

If you'd like to prove this theory to yourself, it's quite simple. Take a golf ball and sand it down until it's perfectly smooth. Then tee off and see how far the ball travels. Then tee off again with a regular ball and see how much farther it goes.

FACTOIDS

The first recorded hole-in-one was hit by Tom Morris, Jr., in the 1868 Open Championship. The oldest golfer to score a hole-in-one was Otto Bucher. In 1985 he hit a hole-in-one at the La Manga golf course in Spain. He was 99 years old.

In 19th-century England, a line was scratched in the dirt as the starting line for race horses. A scratch line was also used in other races. Some competitors started at the line or "started from scratch," while others were given a handicap and could start ahead of the scratch line. That's the origin of the term "scratch," meaning a zero handicap in golf.

The term "golf link" comes from an Old English word *hlinc,* meaning gently undulating land often next to the seashore. The Saint Andrews, Scotland, golf course originally had 22 holes but was later modified to have 18. Because the course ran along the shore, golfers would play nine holes in one direction (called "in") and then play the remaining nine holes (called "out") in the opposite direction. That's where the terms "front nine" and "back nine" came from.

It's interesting that golf spelled backward is flog. For some beginning golfers, flogging might seem better than the humiliation of not being able to hit the ball more than a few yards.

DID YOU KNOW?

Many golfers today become frustrated when they don't hit the ball far enough. If that bothers them, it's a good thing they don't have to play with some of the earlier golf balls.

The very first golf ball was made of wood. It was soon replaced with the "feathery," a ball made by tightly packing goose feathers into a sphere of horse or cow hide. Around 150 years ago, the feathery was replaced by the gutta-percha ball, or "guttie" for short. This ball was made from the rubber-like sap of various tropical trees. The problem was that it was very smooth and didn't travel as far as the rough-surfaced feathery. Strangely enough, the ball seemed to improve with age as it became nicked through constant use.

Realizing the advantage of nicking, manufacturers began putting patterns into the ball such as the "bramble" pattern, which consisted of raised spherical bumps (remember, rough balls travel farther than smooth balls).

Around the turn of the century, a new ball took over the game. It had a one-piece rubber core wrapped in rubber thread and then covered with a gutta sphere. During this time people experimented with different patterns, and someone came up with the dimpled pattern that is still used today.

Manufacturers continued to experiment with new types of balls, using metal, cork, mercury, and other materials for the core. One such experiment used a compressed air core. It sounded good in theory, but there was a minor problem. If the temperature got too hot, the heat expanded the air core, and the ball exploded. As if golfers didn't have enough to worry about.

Why do they pitch baseball overhand and softball underhand? (What about sideways?)

Softball was invented as a way for professional baseball players to practice during the winter. Because they couldn't practice in ice and snow, they decided to play the game indoors. To accommodate the smaller playing area, they used a larger ball and put more players in the outfield.

Because they played indoors, there was no pitching mound. Had they used a mound and pitched in the normal way, the speed of the ball over such a short distance could hurt them. If they didn't use a mound, they would get into bad habits so they would not do as well when playing on a regular playing field. Therefore, it became common for a manager or the coach to simply lob the ball underhand to the batter to get the game started. Eventually the team's pitcher started throwing the ball underhand.

When softball gained popularity in the early 1930s, it became a summer sport as well and was moved outdoors. Because of the smaller field, a mound would have given the pitcher an advantage, so it wasn't used. Pitchers were not allowed to use a windup and had to pitch underhand. In later years, the game of fast-pitch softball became popular and various methods of delivering the ball were invented such as the "windmill" and "slingshot" deliveries.

In professional baseball, much of the speed and control of the ball is determined by the power gained from a combination of the windup, the overhand throw, the movement off the mound, and the follow-through. The same amount of power typically cannot be generated by throwing the ball underhand.

Many people think that a professional pitcher must throw overhand. That's not true. There is no rule that says the pitcher must throw overhand. Although not common, there have been major-league pitchers who threw both sidearm and underhand (called a "submarine" pitch). Juan Marichal, who played for the San Francisco Giants, threw overhand, sidearm, and underhand. In six seasons he won 20 games or more.

The difference between softball and hardball is not just the ball itself. Softball players come in all ages. Unlike baseball, which usually depends on raw strength and athletic ability, softball is more dependent on thinking about what to do and when to do it.

If you're too young, or too old, or too much of a klutz to play hardball, you can probably find a softball game somewhere with teams made up of players who are too young, too old, and too klutzy.

FACTOIDS

Although the first women's softball team was formed in 1895 in Chicago, Illinois, it wasn't until 1996, just over 100 years later, that women's fast-pitch softball debuted in the Olympic Games in Atlanta, Georgia.

The great baseball player Ty Cobb never wore a number on

his uniform. Ty Cobb retired from baseball before the custom of putting numbers on uniforms became a common practice in the late 1920s.

The first major-league baseball player to sign a million-dollar contract was Nolan Ryan in 1979. He signed a four-year contract with the Houston Astros for $4.5 million.

In 1938, Johnny Vander Meer, who played for the Cincinnati Reds, became the first and only pitcher to throw two consecutive no-hit games.

DID YOU KNOW?

There are countless stories about great and near-great baseball players, all usually focusing on their prowess as athletes. Yet there is one great ballplayer few people today have heard of, a player who was not only a tremendous athlete, but who had to overcome adversity that few of us face. His name was William Hoy, but he was known as "Dummy" Hoy.

William Hoy lost his hearing as a young child. In spite of his deafness he became a major-league baseball player, the first deaf player in the game's history. He was small in size, only 5 feet 4 inches, but large in heart. He was a great fielder, and in one game he threw out three base runners at home plate, a feat that has rarely been duplicated.

Because he could not hear the umpire call out balls and strikes, pitchers took advantage of him. So he asked his third-base coach to raise his left arm if the pitch was a ball and his right arm if it was a strike. That's how Hoy knew what had been pitched to him. He also devised the "out" and "safe" signals. All of these signals are still used by umpires in today's games.

In his rookie year, Hoy led the National League in stolen bases. In 1888, he set a fielding record that has not been broken yet. He also hit the first grand slam in the history of the American League.

The next time you watch a baseball game and see an umpire give a hand signal for a ball or strike, or for an out or safe, pause a moment and thank the little man with the big heart for adding so much to the game of baseball.

What are the meanings behind the rings on the Olympic flag . . . both number and colors? (A symbol of unity around the world.)

The Olympic rings, the official emblem of the Olympic Games, are five interlaced rings of blue, yellow, black, green, and red on a white background. The flag was designed by Baron Pierre de Coubertin in 1913. The Olympic rings represent five continents (Africa, the Americas, Asia, Australia, and Europe) and are interlaced to represent the union of these continents. At least one of the five colors is found in the flag of every nation in the world.

The first Olympic flag, which was approximately 10 feet by 6.5 feet, was made at the Bon Marché store in Paris, France, and flew over the Olympic stadium during the 1920 games in Antwerp, Belgium. The Olympic motto *"Citius, Altius, Fortius"* was also on the flag, Latin for "faster, higher, stronger." The flag was made of satin and the rings and motto were embroidered. Because it first flew at the Antwerp Olympics, it was called "the Antwerp flag."

This original flag was flown at every Olympics from 1920 to 1984. After 64 years of use, the flag started showing signs of wear and a new flag made of Korean silk was presented by Korea. It was first flown in the 1988 Olympic Games.

If you look closely at the Olympic flag, you'll see how the

five rings are interlaced so that none can be removed. It's significant that athletes and spectators from all over the world meet at the Olympic games, where they are joined together in a common event.

FACTOIDS

When the modern Olympics started in 1896, a silver medal was given for first-place winners because gold was considered inferior. Gold replaced silver beginning with the 1904 Olympics. Today the gold metals are sterling silver covered with a thin coat of pure gold.

No medals were given in the 1900 Olympics held in Paris, France. Winners were awarded valuable pieces of art.

In 1996, during the Atlanta Olympic Games, Shun Fujimoto of Japan broke his leg during a tumbling run in the floor exercise. Knowing that his team needed him, he decided to compete in the ring competition in spite of the broken leg. He finished with a triple-somersault dismount. He gritted his teeth as pain shot through his leg when he landed. He didn't buckle but stood up. He scored 9.7. With the painful broken leg, Fujimoto had one more thing to do. He managed to climb on top of the podium and stand tall as he was presented with a gold medal.

Today's Olympic creed stresses sportsmanship and the importance of participating rather than winning. In ancient Greece, the original Olympic Games were much different. The only thing that mattered was winning. The victor was crowned with a wreath of olive leaves, while those who finished second and third were sent home in disgrace.

DID YOU KNOW?

There have been many heroes in the long history of the Olympic Games. One such hero competed in the 1936 games and is still an inspiration to athletes today. He was Jesse Owens.

Owens was only 23 when he competed in the 1936 Olympics in Berlin, Germany. He had more to overcome than just athletic competition. Jesse Owens was an African-American. His father was a sharecropper in Alabama, and his grandfather had been a slave. In the 1930s, African-Americans were barred from playing on major league sports teams, a fact rather hard to believe today.

Adolf Hitler, in power in 1936, believed that the German people were the super race and that everyone else was vastly inferior, especially Jews and African-Americans. The stadium in Berlin was to be a showcase for Hitler's belief in the superiority of the German people.

Jesse Owens proved Hitler wrong. Competing against top German athletes, he won four gold medals in the sprints and long jump, a record that was not equaled for almost 50 years when Carl Lewis did the same in 1984. In winning the four medals, Owens broke three Olympic records and tied a fourth. The 10 African-Americans on the United States Olympic team won 13 medals. Hitler refused to shake any of their hands and decided that African-Americans should not be allowed to compete in future games.

One of the most stirring moments occurred when Jesse Owens fouled on his first two attempts at the long jump and had only one jump left. It's been rumored that a German long jumper, Luz Long, told Owens to place a towel behind the takeoff board to use as a starting point. Whether it is true or not, the fact is that Owens's final jump set an Olympic record. Luz Long and Jesse Owens, two competitors, walked off the field arm in arm.

Of all the tributes paid to Jesse Owens during his lifetime, perhaps the most memorable are the words written on his gravestone.

OLYMPIC CHAMPION
1936
A WINNER WHO KNEW THAT WINNING WAS NOT EVERYTHING

Who invented Frisbee? (This one's as easy as pie.)

It all started with a pie. The Frisbie Baking Company of Bridgeport, Connecticut, made pies that were sold to many colleges in New England in the 1940s. The college students soon discovered that if you threw an empty pie tin, it would sail gracefully through the air and could be caught. If you weren't watching when someone threw a pie tin, it could hit you in the head, which wasn't a pleasant feeling. Because Frisbie was stamped on the pie tin, students started yelling "Frisbie" to warn the other students that a metal disc was coming their way.

There is still disagreement among authorities as to whether the company was called the Frisbie Baking Company or the Frisbie Pie Company. There is another controversy because the Frisbie Baking Company also made cookies shipped in round tins: one school of thought insists that students were throwing pie tins, while the other school of thought insists they were throwing cookie tin lids.

In 1948, a Los Angeles building inspector, Walter Frederick "Fred" Morrison, started experimenting with flying discs. At that time people were talking about UFOs and flying saucers, so Morrison thought he might be able to capitalize on the public's interest in flying saucers. He decided to make a flying disc out of plastic. His first creation was called the Arcuate Vane, and another version made out of harder plastic was named the Pipco

Crash. In 1951, he produced his second model, which he called the Pluto Platter.

A new toy company, WHAM-O, liked the Pluto Platter and bought the rights to the design. Unfortunately, it didn't sell very well. The following year, WHAM-O's president, Richard Knerr, when on a marketing trip to the East Coast, discovered Harvard and Yale students playing with pie tins they called Frisbies. He trademarked the name Frisbee in 1959 and sales soared.

Early Frisbees had instructions written on the underside. They were "Play catch. Invent games. To fly, flip away backhand. Flat flip flies straight. Tilted flip curves."

The Frisbee has come a long way since then. There are at least 70 models of flying discs. Ultimate Frisbee, a cross between football, soccer, and basketball, is a recognized sport at hundreds of colleges; there are over 500 Frisbee disc golf courses around the world; and there are championship Frisbee events for dogs and even cats.

It's rather amazing when you realize it all started with an empty pie tin.

FACTOIDS

Fred Morrison's father was the inventor of the automobile sealed-beam headlight.

The outer third of a Frisbee is part of the original patented design and is known as Morrison's slope.

During World War II, Morrison was a prisoner of war in the famous Stalag 13 German prisoner-of-war camp.

The Frisbie Baking Company shut down the same year Morrison received a patent for his flying disc.

In 1968 the U.S. Navy spent almost $400,000 to study the aerodynamics of Frisbees. They put Frisbees in wind tunnels, took films of their flight, and analyzed them with computers. They even built a Frisbee launching machine on top of a cliff. After the money ran out, the research stopped.

The first professional model Frisbee went on sale in 1964.

To help the morale during Operation Desert Shield in 1991, more than 20,000 Frisbees were sent to the troops stationed in Saudi Arabia.

The world's record for throwing a flying disc the farthest is held by Sam Ferrans of La Habra, California, who in 1988 threw a disc a distance of 623 feet, 7 inches.

DID YOU KNOW?

In 1871, William Russell Frisbie was hired to manage a branch of the Olds Baking Company in Bridgeport, Connecticut. He eventually bought the bakery and renamed it the Frisbie Baking Company. William ran the bakery until his death in 1903, and his son managed it until 1940. Other heirs continued to operate the bakery from that time on.

During its peak times, the Frisbie Baking Company opened outlets in other New England states. It grew from a small bakery with 6 routes to a large bakery with 250 routes, and by 1956 the company was producing 80,000 pies a day. When college students started throwing the empty pie tins around, a company spokesperson complained that the bakery had lost around 5,000 reusable pie tins.

The bakery closed its doors in 1958. It's not known where the Yale students purchased their pies after that.

What does "seeded" mean in tennis? (Another popular sport sprouted from it.)

In the late 1800s, lawn tennis tournaments often began with matches between the best players. The worst players were scheduled to play near the end of the tournament. Unfortunately, audiences tended to leave the tournament after the best players had finished.

Tournament promoters realized that they could keep audiences interested if the best players appeared at different times during the tournament. The promoters decided to scatter the top players "like seeds" throughout the scheduled matches so that less expert players would appear in the early matches. The idea of "seeding" is to make sure that the best players meet in the later stages of the tournament.

Although seeding was used in the late 1800s, it wasn't used in the United States National Championships until 1922, and in the Wimbledon Championships until 1924. In 1927, a system was devised that put players in fixed positions depending on merit.

For many years tennis was a sport enjoyed mainly by the more affluent members of society. However, the late 1960s saw a tennis boom. Several factors fueled the new interest in tennis: championships became open to professionals as well as amateurs, television networks began broadcasting tennis matches, and there were changes in fashion and equipment. Tennis balls were made in various colors and tennis rackets became available in many styles and shapes.

Today, tennis is played by millions of people and watched and enjoyed by countless others. Perhaps "seeding" players also helped make the sport more popular.

FACTOIDS

Some people believe that the term "love" for "zero" when scoring a tennis match comes from the French word *l'oeuf,* which means "egg." However, the most accepted explanation is that it comes from "doing something out of love" or "doing it for nothing." In other words, if a person consistently loses, or scores zeros, and keeps playing, that person must truly love the game.

The strange scoring method in tennis was derived from the four quarters on a clock: 15, 30, 45, and 60. Because 60 minutes is the end of an hour, 60 signified the end of the match. Eventually, the 45 was shortened to 40 for some unknown reason.

In medieval times, the French nobility played a game called *jeu de paume,* which means a "palm game," because the ball was hit with an open hand. Later, rackets were used to hit the ball. A player about to serve the ball would yell out *"Tenez!"* (Here!) to let the other player know the game was starting. *Tenez* is the origin of the word tennis.

The first tennis courts were grass and were called lawn courts. The game was referred to as lawn tennis.

The longest match of a Grand Slam tournament was in the 1998 French Open when Spain's Alex Corretja beat Argentina's Hernan Gumy in 5 hours and 30 minutes. The match was only four minutes longer than the previous record set in the 1992 U.S. Open semifinals. That match between Stefan Edberg and Michael Chang lasted 5 hours and 26 minutes, with Edberg finally winning it.

If a player wins all 24 points of a set, it is called a "gold set." The only gold set ever recorded was between Bill Scallon of the United States and Marcos Horever of Brazil. Scallon won by 6 × 2, 6 × 0.

DID YOU KNOW?

Soccer is the most popular sport in the world, but a form of tennis is the second most popular sport. It's table tennis or Ping-Pong. This sport is played by more people in the United States than baseball or football.

Although no one knows for sure, it is believed that table tennis originated in England around the 12th century as a parlor game version of Royal Tennis, as tennis was called then. Initially all the equipment was improvised. A piece of cardboard was the paddle, books were used to form a "net," and the ball was often a ball of string.

In the late 1800s, manufacturers of sporting goods started making official table tennis equipment such as solid rubber or cork balls. These early versions of table tennis were called by various names such as Gossima, Whiff-Whaff, and Flim-Flam.

James Gibb, an Englishman visiting the United States, saw children playing with plastic toy balls and took some back to England to use for table tennis. They were an instant hit. Parker Brothers, Inc., had been making some table tennis equipment at the time and liked the sound of the plastic ball hitting the table. They decided to name their version of the game after the sound of the ball. They called it Ping-Pong.

As the game grew in popularity, national and international associations were formed. The United States Table Tennis Association wanted to purchase rights to the name Ping-Pong but could not come to an agreement with Parker Brothers, so they named the sport table tennis.

You can call the game Ping-Pong, table tennis, Whiff-Waff, or Flim-Flam. However, no matter what you call it, it's an exciting and fun game.

More questions? Try these Web sites.

Olympic Games

http://www.pua.org/olympic.htm

This site has a long list of Web sites devoted to the Olympic Games. It covers both the ancient and modern Olympic Games and includes statistics and other information.

Golf

http://espn.go.com/golfonline/index.html

http://www.worldgolf.com/

Both of these sites contain up-to-the-minute news of the PGA, LPGA, and Seniors tours. They also have pages on golf instruction, equipment, and courses. The World Golf site contains a section on women's golf, as well as classified advertisments.

TENNIS

http://www.tennisserver.com/

This is a comprehensive tennis site with stories and features. It also has daily tennis news sources, equipment tips, rules and codes of tennis, tennis clubs and organizations, and links to other tennis Web sites.

If you are interested in the origin of tennis, the following Royal Tennis Court site provides a history of the original indoor game, plus details of the court and rules. It also has quotes dating back to Chaucer.

http://www.realtennis.gbrit.com/index.htm

PLAYFRISBEE.COM

http://www.playfrisbee.com/

If you or your dog are Frisbee enthusiasts, you'll love this site. It includes information about Ultimate Frisbee, Frisbee golf, freestyle Frisbee, and information about Frisbee-catching dogs. For instance, if you click on "Crucial Information," you'll find a link to a jargon dictionary, a "How to" link which explains how to throw a Frisbee, and a link to the history of the Frisbee.

BOOKS, VIDEOS, AND MERCHANDISE FOR EVERY SPORT

http://www.onlinesports.com/pages/top.html

This is like having a gigantic sports catalog store in your computer. You can look for products by type of sport, supplier, item name, player, or team.

13

Transportation and Travel

Why do diesel truck drivers leave their truck engines running when parked? (Not for a fast getaway.)

Diesel truck drivers keep their engines running for two reasons, even if they're not in the truck. One is that it takes a diesel engine time to warm up. If they turned off the engine, they'd have to wait for it to warm up again before they could start driving.

The most important reason is because of their brakes. Truck brakes work on air pressure, created by a compressor that runs off the engine. If the drivers shut off the engine, air bleeds out of the compressor and the brakes lock up. Sufficient air pressure has to build up before the brakes can be released. If the engine is kept running, the compressor keeps the pressure up, and the driver can release the brakes as soon as he gets into the truck.

A typical semi truck weighs 20 to 30 times as much as an automobile. Even with the best brakes, it cannot stop as fast as an automobile. In fact, a semi has a stopping distance 20 to 40 percent longer than a passenger car. When a big rig and an automobile collide, it's no contest. In fatal crashes involving an automobile and a large truck, 98 percent of the people killed were drivers or passengers in the automobiles.

It's a good idea to remember that statistic the next time you think about swerving in front of a truck to get ahead. You might even have the right of way. But that won't help you if the truck can't stop, and being dead right isn't always a good thing.

FACTOIDS

Over half a million orange barrels used for highway construction zones are produced each year.

In Massachusetts it's against the law to deliver diapers on Sunday, even in an emergency.

In spite of what some people think, wind tunnel tests have proven that if you drive a pickup truck with the tailgate up, there is less drag and you get better gas mileage than if you leave the tailgate down.

Truck stops pump over 13 billion gallons of diesel fuel a year.

DID YOU KNOW?

The Terex Corporation manufactures off-highway mining trucks. One of its trucks, called the Titan, is used at the Elkview Coal Company in Sparwood, British Columbia. It is the world's largest dump truck.

The truck is over 22 feet high, almost 66 feet long, and

almost 25 feet wide. It weighs 260 tons but can carry 350 tons. When the box is raised to dump a load, the truck is 56 feet high.

The truck has 10 tires, each having a diameter of 11 feet. If you were six feet tall and stood next to the tire, your head would just be past the middle of the tire.

The 16-cylinder engine generates 3,300 horsepower. The fuel tank holds 1,300 gallons. To give you an idea of how much fuel that is, assume you have a 20-gallon tank in your car and use a full tank each week. At that rate, you would only use 1,040 gallons for the entire year, not enough to fill up the Titan's tank just one time.

Heavy-duty mining trucks like the Titan take time to build and are expensive. In order to be cost effective, they must operate for much longer hours than a normal truck. There are 8,760 hours in a year, and a typical mining truck is operated 5,000 or 6,000 hours a year. That means it runs 14 to 16 hours a day, every day of the year.

Another giant mining truck built by the Terex Corporation set a new record for continuous operation at a mine on Minnesota's Iron Range. It was continuously operated for 8,128 hours. In other words, it worked for well over 20 hours a day, 7 days a week, for a full year.

If you ran your personal automobile 20 hours a day, 7 days a week, do you think it would last for a year?

If all the trains in the country were put end to end, how long would the line be? (It doesn't include the Cannonball Express.)

If all the freight-hauling railroad trains in the United States were lined up end to end, they would form a line over 18,000 miles long. This doesn't include passenger, commuter, or excursion trains. Even so, there wouldn't be enough room in this country to keep them in a straight line. The line would run coast to coast six times.

The following facts were used to calculate the length of the line:

1. There are 534 freight railroad companies in the United States.
2. The freight railroad companies have approximately 23,500 locomotives.
3. The average number of cars per train is around 68.
4. The average length of a car is 60 feet.

If you multiply the length of each car (60) times the number of cars in a typical train (68) times the number of freight trains in the United States (23,500), you'll get an answer of 95,880,000 feet. Converting this to miles gives you 18,159 miles.

FACTOIDS

The first rail was laid in the United States in 1828, but after two years it still consisted of just 30 miles of track.

Andrew Jackson was the first U.S. president to ride the train when he traveled from Baltimore, Maryland, to Ellicott's Mills in 1833.

In Western Australia, on one section of track, trains travel almost 300 miles in a straight line without making a single turn.

In 1850, the United States had over 9,000 miles of track in operation, which was more than all of the track in the rest of the world combined. In just 10 years the figure more than tripled to 30,000 miles of operating track.

Prior to railroads, clocks in cities within the same state often varied by 30 minutes or more. The railroad created time zones with Standard Time in 1883, so that train schedules could be met.

In 1944, a train in Salerno, Italy, stalled in a long tunnel, but the engine continued to burn, producing fumes from a lower-grade coal used during the war. By the time the crew got the train moving again, 500 people had died from carbon monoxide poisoning.

DID YOU KNOW?

Many people have heard the song about the brave engineer Casey Jones, but not many know the true story of what happened the night Casey rode to his death.

Casey's real name was John Luther Jones, and he was born near the town of Cayce, Kentucky. People pronounced the name of the town as "Kay-see," and that's how he got the nickname Casey.

His trademark was the way he blew the train's whistle. When people heard the telltale signal, they'd say, "There goes Casey Jones." In 1900 he was transferred to Mississippi and became an engineer on the Illinois Central famous "Cannonball Express."

On Sunday night, April 30, 1900, one of the engineers became ill and Casey volunteered to take his run. He slowly pulled engine No. 382 out of the yard into a foggy night. There was a light rain. Although starting behind schedule, Casey was determined to make up the time. He was pulling only six passenger cars and thought he could make the express run in record time.

Casey rounded a curve going about 75 mph. Up ahead two freight trains had pulled off on a siding to let the express pass. Unfortunately, an air hose had broken on one of the cars and the last two cars hadn't reached the siding. The fireman yelled out that there was a caboose sticking out on the track. Casey applied the air brakes and managed to slow the engine to about 50 mph. The fireman realized the train couldn't stop and jumped from the engine.

The Cannonball Express plowed into the freight train's boxcars, jumped the track, and kept going for some distance before flipping over on its side. Casey Jones was killed. He could have saved himself by jumping, but he stayed and continued applying the brake, thereby lessening the impact. Because of this, none of the passengers was killed or seriously injured.

The freight railroad company claimed that Casey was completely at fault because he had ignored warning signals given by a flagman from the freight train.

In the history of railroads, this was not a major accident. However, an engine wiper, Wallace Saunders, idolized Casey and wrote a song about the wreck of the Cannonball Express. The song became popular throughout the country and made Casey Jones an American folk hero.

Casey Jones was just one of a long line of brave locomotive engineers who died with one hand on the whistle and the other on the air brake.

What is the fastest propeller-driven plane in the world? (Hope you can bear this one.)

The world's fastest propeller-driven airplane was the Russian TU-95/142 *Bear.* The plane had four engines, each producing 14,795 horsepower. Each engine drove an 8-bladed counter-rotating propeller. During level flight, the airplane could reach speeds up to 574 mph.

The fastest speed of any piston-engine aircraft was attained by *Rare Bear,* a modified Grumman F8F Bearcat. In 1989, Lyle

Shelton flew it over a timed course in Las Vegas, Nevada, and reached a speed of just over 528 mph.

It's interesting that both record-breaking airplanes were named Bear.

FACTOIDS

The world's smallest piloted biplane was *Bumble Bee II,* built and flown by Robert H. Starr. It was only 8 feet 10 inches long, with a wingspan of just 5.5 feet, and could reach speeds of up to 190 mph. It crashed and was destroyed in 1988. The smallest monoplane is *Baby Bird,* built by Donald Stilts. Just 11 feet long, with a wingspan of 6 feet 4 inches, it has reached speeds of 110 mph.

In 1998, pilot Brian Milton and copilot Keith Reynolds were the first to fly a microlight airplane around the world. They also broke a 74-year-old record for global flight in a single-engine, open-cockpit aircraft.

Although most people know that Charles Lindbergh made the first solo nonstop transatlantic flight, not that many know that the first nonstop transpacific flight was made by Major Clyde Pangbonn and Hugh Herndon when they flew from Sabishiro Beach, Japan, to Wenatchee, Washington.

The largest propeller ever used on an airplane was 22 feet 7.5 inches in diameter (taller than three six-foot men standing on top of one another). It was used on a Linke-Hofmann R II built in Germany and flown in 1910.

DID YOU KNOW?

An aircraft race held in 1934 is considered to be "the great air race," although it was more like a free-for-all.

The city of Melbourne, Australia, decided to host an airplane race between their city and London, England, in order to celebrate the centenary of the state of Victoria. A millionaire agreed to sponsor the race if they named it after him, which they did. It was called the MacRobertson England to Australia Air Race.

The race consisted of six legs, the first four more than 2,000 miles each. The course covered 16 countries and 3 continents, and required day and night flying over the snake-infested jungles of India, jagged mountains, deserts, the Bay of Bengal (which nobody had ever flown over), and the shark-infested Timor Sea. There was also a good chance the pilots would have to brave fierce tropical storms and blinding sandstorms that could rise to 20,000 feet.

There were no rules except the flight path. An airplane of any size with any amount of power could enter, and the crew could be as small or as large as desired. The race started in London at dawn on October 20, 1934. Although 64 people registered for the race, only 20 showed up.

In the first two days some pilots became lost, others had mechanical problems, and the first accident occurred when a plane flipped over when landing in Syria.

On the third day the first fatality occurred. A plane with two Englishmen crashed and both were killed. That same day, two other Englishmen, Charles W. A. Scott and Tom Campbell Black, flew the last 300 miles over water with just one engine functioning. They risked death when they landed on a rain-soaked field to refuel.

Scott and Black had had only two hours sleep since leaving London. They triumphantly landed their deHaviland DH88 Comet in Melbourne just 70 hours, 54 minutes, and 18 seconds after leaving London. They had flown halfway around the world in just under three days.

Entries from the United States finished second and third.

Hail Britannia!

What was the *China Clipper*? (It was definitely not an oriental barber.)

In 1934, no one believed it would be possible to fly passengers across the Pacific Ocean. In fact, 13 pilots had already died attempting the perilous journey. However, the founder of Pan Am, Juan Trippe, believed that an airplane could fly nonstop from San Francisco to Honolulu and then refuel by hopping from island to island until it reached the Philippines, 8,000 miles away. Trippe's plan was to establish transpacific mail service.

The islands to be used for refueling stops needed airbases, so Trippe loaded ships with the material needed to build the bases and sent them to Midway Island, Wake Island, and Guam.

The airplane selected to make the trip was a giant flying boat, the Martin M-130, named the *China Clipper*. It was the largest flying boat ever built up to that time, with a wing span of 130 feet and a length of 84.5 feet. In 1935, it took off on its maiden flight with Captain Ed Musick at the controls. It lifted off the waters of San Francisco Bay and flew the 8,000 miles to Manila, in the Philippines, in 59.75 hours flying time. It carried 100,000 pieces of mail on the first transpacific airline flight. A week later, the *China Clipper* returned to San Francisco carrying 108,000 pieces of mail.

The momentous flight made the world seem a little smaller at that time. Two more Martin aircraft were put into service: the *Philippine Clipper* and the *Hawaii Clipper*. In 1936, the *Hawaii Clipper* began the first passenger service between California and the Philippines. The plane could hold 32 passengers and had overnight berths for 18. Among the nine passengers on the first flight were Fortune Ryan II, the man who had built Charles Lindbergh's plane, *Spirit of Saint Louis*.

Pan Am's clipper ships continued flying to the Orient until the outbreak of World War II, with only one major tragedy.

Captain Ed Musick, the first to fly the transpacific route, and his five crewmen died when the *Samoan Clipper* disappeared on a flight from Pago Pago to New Zealand.

FACTOIDS

When the *China Clipper* took off on its maiden flight, it was supposed to fly between the towers of the uncompleted Bay Bridge. However, the pilot realized he couldn't get up enough speed to clear the wires, so he dove down at the last minute, flew under the bridge cables, and slithered his way through construction wires until he was in the clear.

Later models of the Pan Am clipper fleet had dressing rooms, a dining salon, a bridal suite, and seats that converted into bunks. In 1943, President Franklin D. Roosevelt celebrated his birthday in the salon of a Pan Am clipper on his way home from a conference with Winston Churchill in Casablanca.

When Japanese planes bombed Wake Island, they strafed the *Philippine Clipper* sitting in the lagoon. Although more than 90 bullets pierced the aircraft's hull, it could still fly. The pilot flew it just above sea level to avoid detection. After refueling at Midway Island, the clipper flew on to a safe landing in Hawaii.

The *Pacific Clipper* was halfway between New Caledonia and New Zealand when Captain Robert Ford heard that war had broken out in the Pacific. Rather than risk crossing the Pacific Ocean to reach San Francisco, he decided to go the other direction and circle the globe. When he arrived in New York a month later, he had not only flown 31,000 miles, but had also completed the first around-the-world flight by a commercial airliner.

DID YOU KNOW?

The term "clipper" is probably derived from the verb "clip," meaning to move quickly. It was given to sleek sailing ships known for their beauty and speed. Such ships were called Yankee clippers. The age of the clippers lasted for just a decade, from the late 1840s to the mid 1850s. Nonetheless, they had a profound impact on world trade.

Prior to the design of the Yankee clipper, England ruled the seas with heavily armed merchant ships that had a top speed of three or four knots. However, the long, lean, and beautiful Yankee clippers could travel at 20 knots or more. Even when heavily loaded, they could maintain such speeds over long periods of time. It was not uncommon for a clipper ship to average 400 miles in a 24-hour period.

Sadly, most of the sleek clipper ships have long since gone. A few, like the *Cutty Sark,* are now museums.

Although some of these tall ships still sail the seas, they are usually not available to the public. One exception is the *Clipper City,* a perfect replica of the original clipper.

Pan Am's clipper aircraft have been gone for decades. But if you want to sail on a sleek clipper ship, you can charter the *Clipper City* at its home port in Chesapeake Bay, Virginia.

What makes sled dogs run? (They're not running to find mush to eat.)

There is one main reason why sled dogs run. They love it! It's almost an inbred trait. Although they all love to run, they have to be trained to run harmoniously as part of a dog team. How well they learn to perform in a team depends on their driver, who is called a "musher."

Although almost any large dog can be trained to pull a sled, the two most popular breeds are the Alaskan malamute and the Siberian husky. The Alaskan malamute is a large dog, weighing 75 to 85 pounds, and is used for pulling heavy weights. The Siberian husky is faster and smaller, weighing 35 to 65 pounds. The huskies were originally used for herding reindeer, pulling loads, and other tasks. Other popular sled dogs are the samoyed, laika, and American Eskimo breeds. They are all large and powerful dogs with thick coats and great stamina. A team of such dogs can pull a sled and person for hundreds of miles.

The use of sled dogs goes back thousands of years. They were used not only for transportation but also for protection, hunting, and companionship. Famous polar explorers such as Byrd, Peary, and Amundsen all used sled dogs. Around 1873, the Royal Canadian Mounted Police used dog-team patrols, and other dog teams delivered mail throughout Alaska and Canada.

In some areas of the world, dog teams are still used for transportation, but most of them are now engaged in the sport of sled dog racing.

FACTOIDS

Siberian huskies were raised in Siberia by the Chukchi people, who kept the breed pure for hundreds of years. When pulling a sled, a husky can burn twice the number of calories burned by an athlete cycling in the Tour de France bicycle race.

Robert Peary was the first man to reach the North Pole. He took 240 king Eskimo dogs with him, and 40 of them went all the way to the pole. Peary might not have reached the pole without the help of what many consider to be the best sled dogs in the world.

A dog sled race was one of the events in the 1932 Winter Olympics. The race was won by E. Goddard of Canada.

Although dog sled drivers are called mushers, they do not

start the team moving by yelling "mush" (a corruption of the French word *marche,* meaning "go"). They use the words *hike* to start the team, *gee* to turn it right, *haw* to turn it left, and *easy* to slow it down.

DID YOU KNOW?

In 1925 a case of diphtheria was discovered in Nome, Alaska. To avoid a disastrous epidemic, the residents had to be inoculated with antitoxin serum, but the only available serum was in Anchorage. The only two planes in Fairbanks had been dismantled for the winter and even had they been available, flying an open cockpit plane in –40°F stormy weather would have been almost impossible. If a plane carrying the serum crashed, the valuable serum would be lost.

It was thought that the only solution was to send the serum from Anchorage to the small town of Nenana by train and then relay it overland by dog sleds to Nome, 674 miles away. To make the run, 20 drivers were recruited, along with 100 dogs. The drivers were Native Americans, Inuits, mail carriers, and freighters who all had experience traveling by dog sled.

The first team set out with the precious 20-pound package of serum. The package was relayed from team to team as the dog sled teams drove through Alaska's harsh interior and across the frozen ice of the Bering Sea, enduring Arctic blizzards along the way. In just a little over five days (127 hours), the last team, led by a dog named Balto, swept into Nome with the serum. Because of the determination of the drivers and the heroism of their dogs, many lives were saved.

Since 1973, this historic serum run has been commemorated with the Iditarod Sled Dog Race between Anchorage and Nome, Alaska. The course is around 1,100 miles long and crosses two mountain ranges as well as the frozen pack ice of Norton Sound. It takes two to three weeks to complete the race.

Early Athabascan Indians in Alaska called their hunting ground *Haiditarod,* meaning "the distant place." Miners who settled on the old hunting grounds spelled it as Iditarod, and that's what the race is called today.

The Iditarod race is not the only thing that commemorates the heroic serum run of 1925. There is also a statue of Balto, the dog that led the final dog team into Nome, in New York City's Central Park. The statue is dedicated to the sled dogs and describes their heroic achievement. The last three words of the inscription aptly describe the brave sled dogs of the frozen North: "Endurance. Fidelity. Intelligence."

Who holds the record for walking the farthest? (He's still walking.)

Arthur Blissitt of North Fort Myers, Florida, has walked 33,151 miles. Not only has he walked farther than any other person in the world, he has carried a 12-foot, 40-pound wooden cross with him on the entire journey.

Blissitt operated a coffee shop on the Sunset Strip in Los Angeles, California. One day he decided to place a 12-foot-high cross on the wall. Many patrons were drawn to it and one evening he said that God told him to take the cross to the people.

On Christmas Day, 1969, Blissitt started his pilgrimage from the Sunset Strip. He started walking, and he walked and walked. He has carried the cross through 277 countries, 49 of them at war, over all 7 continents, to the Dead Sea, to the top of Mount Fuji in Japan, and to the depths of Carlsbad Caverns in New Mexico. He carried the cross through frozen Antarctica, across the steaming Darien jungle from Panama to Colombia, and over the hot deserts of Turkey and Egypt. He even carried the cross across the Panama and Suez canals.

Although Blissitt's pilgrimage was to spread the word of

Jesus, not everyone has been cordial to him. He has been arrested 24 times and once faced a firing squad.

On Christmas Day, 1998, Arthur Blissitt returned to the place where he had started 29 years before. It had taken him almost 30 years to complete his incredible journey.

Blissitt has not stopped walking, however. He now plans to walk through the states that he missed on his original journey.

Keep on trucking, Arthur!

FACTOIDS

Blissett has walked with a crowd of 70,000 people across Poland.

Pope John Paul II welcomed Blissett to Rome, Yassar Arafat welcomed him to Beirut, and Prime Minister Menachem Begin of Israel let him sleep at his house.

The Roman emperor Hadrian toured his entire empire on foot. He marched 21 miles a day while wearing full armor.

Because Blissett wears out a pair of shoes on an average of every 500 miles, he has gone through over 66,000 pairs of shoes.

In 1809, Scottish Captain Robert Barclay walked 1,000 miles in 1,000 hours. It is rumored that over 100,000 people bet on whether he could do it or not.

Although he had to eat while on his journey through more than 250 countries, Blissett wasn't particularly fond of some of the food he was given, including squid in ink, monkey leg, and rat soup.

In 1861, Edward Payson Weston walked 1,326 miles from Portland, Maine, to Chicago, Illinois, in 25 days.

DID YOU KNOW?

Although walking exercises your legs, so does bicycling. In 1884, Thomas Stevens of San Francisco, California, certainly got his exercise while bicycling. His bicycle had a front wheel that was 50 inches high and a back wheel that was 16 inches high. There was no chain and the pedals were attached to the front wheel. Stevens hopped on his bicycle and took off on a very long trip—around the world. It took him 2 years, 8 months, and 13 days. When he got back home, he had ridden his bicycle for 13,500 miles.

The longest bicycle trip around the world was made by Fred Birchmore in 1935–1936. It took him just a year to complete his 25,000-mile journey. Birchmore left from New York, working his way on a freighter bound for Rotterdam, Holland. Some days later, he bought a bicycle in Gotha, Germany, and started his trip, which took him through 40 countries. He named his bicycle Bucephalus, the name of Alexander the Great's white horse.

Birchmore's trip was far from a leisurely ride through the countryside. He was almost killed by an avalanche on Switzerland's Matterhorn, was attacked by an Arab, and had to avoid crocodiles in the Nile. He was also chased by a pair of king cobras as well as stalked by a tiger in the jungles of Southeast Asia. Disease was also an enemy. Birchmore had an attack of malaria when he was alone in the jungle. Although burning with fever and half delirious, he kept fighting his way through the jungle until he found a hospital in Siam.

Fred Birchmore was 24 when he started his bicycle trek across the world. That was only the start. Four years later he made a 12,000-mile trip around North America. When he was 62 years old, he bicycled across 10 European countries. When he was 70, he completed a 300-mile rock-climbing expedition, and when he was 74 he made a balloon trip to Switzerland. In 1996, at age 85, Birchmore carried the Olympic Torch in Atlanta.

Both Thomas Stevens and Fred Birchmore wrote books about their adventures. Each used the same title, *Around the World on a Bicycle.* That's fine. They both deserve to use that title for books about their astounding trips.

Oh, yes, Bucephalus, Birchmore's bicycle, is now resting comfortably in the Smithsonian Institution.

More questions? Try these Web sites.

IDITAROD

http://www.dogsled.com/

Everything you want to know about Alaska's famous Iditarod dog sled race. It has the history of the race, race rules, a trail map, race trivia, and up-to-date news stories. It also has links to other mushing sites in case you're interested in dog sled racing.

TRAINS

http://www.trainWeb.com/

This site is definitely for the railroad enthusiast. Major sections include rail travel, rail industry, model railroading, live Webcams, photos, and a section for children.

If you click on "Rail travel," you'll see a long list of related travel sites, including route guides, tourist railways, Amtrak schedule and stations, and rail excursions.

BICYCLE HISTORY

http://members.tripod.com/KatNJaz/index%2D5.html
http://www.speedplay.com/whatshappening/pedhistory.html
http://www.ctuc.asn.au/bicycle

These three Web sites give you a varying view of bicycle history, including a sketch of a bicycle by Leonardo da Vinci, and the 1790 running machine, which was nothing more than two wheels connected by a beam. It was propelled by being pushed with the feet on the ground, like a scooter.

Road Construction Information

http://www.randmcnally.com/plan/road%5Fconstruction.ehtml

Even if you've carefully planned your trip, unexpected roadway construction can cause frustration and delays. This site can solve that problem. Select a state, highway, and date and you'll see a list of all active road construction, including the type of construction, when it started, and when it will be finished.

Calculating Distances Between Cities

http://www.indo.com/distance/

You can use this calculator to find the latitude and longitude of major cities around the world, as well as the distance between any two cities. If you enter the name of just one city, the latitude and longitude of that city is shown. If you enter the names of two cities, the site displays the distance between the two cities, as well as each city's latitude, longitude, elevation, and population.

You can click on one link to see a map showing both cities, and you can click on another link to obtain driving directions from one city to the other.

14

United States

How many people died during the flu epidemic of 1918? (Deadlier than World War I.)

In 1918 a soldier in Kansas was admitted to a military base hospital complaining of flu-like symptoms. Doctors discovered that he had a strain of flu known as Spanish influenza. The name came from an erroneous belief that the strain had originated in Spain.

The flu took just seven days to blanket the United States, and by the end of the year, 675,000 people, or roughly 2.5 percent of the country's population, had died from it. The number of people who died from flu in that one year exceeded the number of U.S. citizens who died in all the wars in the 20th century. In the month of October alone 200,000 people died.

The flu epidemic was not confined to the United States. It spread worldwide in just three months, and between 25 and 40 million people died during the epidemic, more than quadruple all the casualties in World War I. The flu epidemic spread faster than any virus in history and killed more people in less time than all of the great plagues of the past.

Some locations suffered more than others. In Alaska, 60 percent of the Inuit population died. In one village alone, 85 percent of the inhabitants died within a week.

The Spanish flu was a deadly mutation of the flu strain. When an epidemic like this hits, no vaccine can help, because the disease spreads around the world faster than any laboratory can find an effective vaccine.

Few people today have ever heard of the Spanish influenza epidemic. It is rarely mentioned, perhaps because people living at the time were preoccupied with World War I, or because people simply tend to forget tragic events.

Still, it's amazing that an epidemic that killed almost 40 million people in this century is basically forgotten.

FACTOIDS

"Influenza" is derived from the medieval Latin word *influentia,* meaning influence, from the belief that epidemics were due to the influence of the stars.

It is estimated that since the 1700s, global flu epidemics have occurred every 10 to 30 years. The two most recent outbreaks occurred in 1957 and 1968, which means we might be overdue for another major flu epidemic.

A person with the flu normally recovers within a week or two. However, sometimes the flu can cause life-threatening complications such as pneumonia. In an average year, the flu is responsible for approximately 20,000 deaths in this country.

Scientists have taken flu virus samples from the frozen body of a man buried in 1918 in a remote Alaskan village. They hope the samples will help them learn how to prevent a similar outbreak in the future.

In 1957–1958, 70,000 Americans died from Asian flu; 34,000 died in 1968–1969 from the Hong Kong flu.

DID YOU KNOW?

If you have nausea, vomiting, and/or diarrhea, you probably don't have the flu. These symptoms are rare in the flu. The so-called stomach flu is usually due to a mild form of food poisoning or other gastrointestinal illness. In fact, many people don't understand the differences among flu, colds, and food poisoning.

In general, if you are nauseated and vomiting a lot, you probably have food poisoning rather than the flu.

Here's how to tell the difference between the flu and the common cold.

Symptom	Cold	Flu
Fever	No	High (102–104°F. May last for a few days or more.)
Headache	No	Yes
Aches and pains	Slight	Severe
Fatigue	Mild	Lasts for weeks
Exhaustion	No	Yes
Stuffy nose	Usually	Sometimes
Sneezing	Usually	Sometimes

Symptom	Cold	Flu
Sore Throat	Common	Sometimes
Cough or chest discomfort	Mild, hacking cough	Common, often quite severe
Other	Sinus congestion or earache	Bronchitis or pneumonia, possibly life threatening

Although disease and death were prevalent during the Spanish flu epidemic of 1918, people did not lose their sense of humor. The following poem was recited throughout the country.

I had a little bird,
Its name was Enza.
I opened the window,
And in-flu-enza.

Is it true that President Zachary Taylor was poisoned? (Arsenic-flavored cherries, perhaps.)

Zachary Taylor, nicknamed "Old Rough and Ready," had been president for only 16 months when he died. On July Fourth, 1850, he ate fresh cherries, raw vegetables, and buttermilk, then participated in ceremonies in the blistering heat. He died five days later, the second president to die in office (William Henry Harrison was the first). Some say he died of sunstroke, some say he died by eating contaminated food. No one knows for sure exactly what killed him or how he became ill.

Some historians believe that President Taylor did not die of gastroenteritis or sunstroke, but was actually poisoned by supporters of Vice-President Millard Fillmore who wanted him to be president.

To settle the matter once and for all, Zachary Taylor's body was exhumed in 1991, 141 years after his death. After a thorough examination by the Kentucky state medical examiner, it was proven that he had not been poisoned.

Zachary Taylor was not the first president to be exhumed. In 1900, Abraham Lincoln's tomb had to be torn down and his coffin was placed in a temporary grave. Because Lincoln's remains had been moved 17 times since the original burial, there were many rumors that the coffin did not contain the body of Lincoln. After much controversy, it was decided to open the coffin. All 23 witnesses agreed without doubt that it was the body of Abraham Lincoln. Today Abraham Lincoln lies buried 10 feet beneath the floor of the Lincoln Memorial, his coffin encased in a solid block of cement.

Distant relatives of Lincoln's assassin, John Wilkes Booth, claim that Booth escaped after the assassination and lived for years under assumed names. They believe that someone else was buried in his grave and have wanted the body exhumed to prove their theory. The courts have refused to allow exhumation of the body.

FACTOIDS

In a meeting with southern leaders who threatened secession, President Taylor told them that he would personally lead the Union Army to enforce the laws of the land. Ironically, 11 years later his only son served as a general in the Confederate Army.

Taylor was opposed to the expansion of slavery, yet he owned more than 100 slaves.

The first presidential election held in all the states at the same time was in 1849, when Zachary Taylor was elected president.

Although he won major victories as a soldier, he had short legs and needed help when mounting his horse.

At Dolly Madison's funeral in 1849, Zachary Taylor coined the term "first lady" when he gallantly said, "She will never be forgotten, because she was truly our first lady for a half century."

Taylor never registered to vote and didn't even vote in his own election.

Because President James Polk ended his term on Saturday, Zachary Taylor should have been sworn in on Sunday. However, he refused to take his oath of office on a Sunday and wasn't sworn in until Monday. The United States didn't have a president for one day in 1849.

DID YOU KNOW?

In 1849, a town in California was founded by miners from Wisconsin. In honor of General Zachary Taylor, they named the town Rough and Ready.

A year later, the miners decided they didn't like the laws made by outsiders in the nation's capital, 3,000 miles away. They held a town meeting and decided to secede from the Union. They drew up articles of secession and established the Republic of Rough and Ready.

The republic was short lived. When the miners started preparing for their July Fourth celebration, they realized that they were now an independent country. Since they were no longer part of the United States, they had no reason to celebrate. Something had to be done.

In a hastily called election, the town overwhelmingly voted to rejoin the Union and become part of the United States once more. It's been said that the celebration the town held that year was the most riotous July Fourth celebration ever held anywhere in the country.

You can visit Rough and Ready today. It's in Northern California in Nevada County, not far from the town of Grass Valley.

Fortunately, or unfortunately, depending on how you look at it, today's July Fourth celebrations are no longer the most rowdy and wild in the country.

Which state has the most shoreline? (Water, water, everywhere.)

A good guess might be California, because one side of the state borders the ocean for 840 miles. Another guess might be Hawaii, because it's surrounded by water. Or it could be Florida, which has the ocean on three sides. However, none of these is the correct answer.

Minnesota has more shoreline than California, Hawaii, and Florida combined. Although the Minnesota license plate proudly proclaims, "Land of 10,000 Lakes," there are actually 11,842 lakes. About 10,000 years ago the state was covered with glaciers, some up to a mile deep. As the glaciers moved and eventually melted, they gouged the land and created thousands of rivers, lakes, and ponds.

Although Minnesota has almost 12,000 lakes, Wisconsin has 14,000 lakes, and Alaska is the champion with 3 million lakes.

However, if all the rivers, lakes, and ponds are included, then the Minnesota shoreline is 93,000 miles, more than any other state. If Alaska ever measured the shoreline of all of its lakes, it would probably win. Alaska's estimated ocean shoreline, including islands and inlets, is 47,300 miles. Until Alaska can pull together a shoreline estimate for its lakes, Minnesota remains the shoreline king.

Part of Minnesota's shoreline is on Lake Superior, the second largest lake in the world, covering some 31,700 miles. Only the Caspian Sea is larger at 144,000 square miles. However, Lake Superior is the largest freshwater lake in the world.

Rivers in Minnesota flow in three directions: north to Hudson Bay, east to the Atlantic Ocean, and south to the Gulf of Mexico.

The name Minnesota comes from *minisota,* which means

"sky-tinted waters." The state is sometimes called the "land of sky-blue waters," which is appropriate for the state with the most shoreline in the United States.

FACTOIDS

Yellowstone Park's Yellowstone Lake is the highest lake in the country at 7,735 feet above sea level.

The largest man-made lake in the world is at the Owen Falls dam in Uganda. Hoover Dam in the United States is number 27 on the list of largest man-made lakes.

Crater Lake, Oregon, is the deepest lake in the United States. It is 1,932 feet deep. The deepest lake in the world is over one mile deep. Siberia's Lake Baikal, has a maximum known depth of 5,371 feet.

Oklahoma has the most man-made lakes in the United States.

The Lost Sea lake, discovered in 1905, is the largest underground lake in the country. It is 300 feet below the surface in Craighead Caverns in Sweetwater, Tennessee.

The shortest river in the country is the D River, near Lincoln City, Oregon. It connects Devil's Lake to the Pacific Ocean and is only 120 feet long, not even half the length of a football field.

DID YOU KNOW?

Shoreline may be beautiful to look at or walk along, but it can be deadly for ships if night or fog makes the shoreline difficult to see. Many shorelines are littered with the wrecks of unfortunate ships.

Lighthouses are positioned along the coasts of oceans,

inland seas, and lakes to warn ships of dangerous reefs and shoals. The first lighthouse built in the United States was the Boston Light, erected in 1716.

One of the most famous lighthouses is the Whitefish Point Light, which guards the treacherous southeastern end of Lake Superior, known as the "Graveyard of the Lakes." In spite of the lighthouse, there have been 70 major shipwrecks in that area.

The Whitefish Point Light has shined faithfully for over 150 years. That is, except for one fateful night when the light was blacked out. The light was not all that was lost that night.

The *Edmund Fitzgerald* was the world's largest freshwater freighter when it was launched in 1958. The 729-foot ship could easily have held 350,000 people. It plied the waters of Lake Superior, stopping at port cities such as Duluth and Toledo.

On November 10, 1978, the 20-year old ship was in the midst of a fierce storm, fighting 80-mph winds and 25-foot waves. It is believed that the captain was heading for Whitefish Point to find calmer waters. But that night the Whitefish Point Light was not shining. The furious gale had broken power poles and there was no electricity to power the light. Some people believe that had the light been working, the *Edmund Fitzgerald* might have reached calmer waters and been saved, but no one knows for sure.

Suddenly the *Edmund Fitzgerald* was hit by two tremendous waves that broke the ship in half. It sank in just 10 seconds. All 29 crew members were lost.

The bell of the ship was recovered in 1995. Since then a ceremony has been held each year at Whitefish Point to honor the crew of the *Edmund Fitzgerald*. The bell is tolled 29 times, once for each of the crew members. It is then tolled one more time in memory of all sailors lost on the Great Lakes.

Why does the Great Seal of the United States have seven white stripes and six red stripes, when it's the opposite on the flag? (No matter what the color, it still comes out to be 13.)

The flag has red stripes on the outer edges so that it can be seen better. On the shield in the Great Seal, the outer stripes are white so the shield can be seen better against the gold background of the seal. In both cases, there are 13 stripes to represent the original 13 states.

Virtually all countries have government seals to authenticate important international documents. The founders of the United States knew that the new country needed both a seal and a national coat of arms that would be a symbol of the country.

On July 4, 1776, Benjamin Franklin, John Adams, and Thomas Jefferson were appointed by the Continental Congress to design a seal for the United States. The task wasn't as simple as it appeared. The seal had to use just a few symbols and words to depict what the United States was at that time and what it would be in the future.

It wasn't until six years later that the final design was approved. William Barton created the design, which shows an American bald eagle holding a ribbon in its mouth with the Latin words *E pluribus unum*, which mean "one out of many." The eagle holds the arrows of war with one talon and the olive branch of peace in the other. The reverse of the seal displays an unfinished pyramid with an eye above it. The eye represents the eye of Providence. Most representations of the seal, such as that on the one-dollar bill, show both the front and reverse of the seal.

The seal is an engraved metal die that impresses its design into a document, similar to the embossing of a notary public's seal. It is used to authenticate treaties and international agreements, and appointments of ambassadors and other foreign service officials, as well as other important government documents.

When the seal is used, a blank paper disc is first glued to the document. Then the paper is put into a press with the die and counterdie. When the press is closed, it creates the image of the front of the seal on the paper disc.

In 1885, Congress allocated money to cut a die for the reverse of the seal. To this day that die has never been made, and the only die is for the front of the Great Seal.

FACTOIDS

The Great Seal of the United States is the only government seal in the world with a design on both sides.

The bicentennial of the Great Seal occurred in 1982, the same year that marked George Washington's 250th birthday and Franklin Delano Roosevelt's 100th birthday.

The number 13, indicating the 13 original states, is used throughout the Great Seal. For example, there are 13 stars, 13 stripes in the shield, 13 arrows in the eagle's left talon, 13 olives and leaves in the right talon, and 13 layers in the pyramid. The inscriptions *E Pluribus Unum* and *Annuit Coeptis* each consist of 13 letters.

On the first Great Seal created in 1782, the eagle looked more like a turkey.

It is believed that the first illustration of the reverse of the Great Seal was published in the October 1786 issue of *Columbian Magazine,* published in Philadelphia.

DID YOU KNOW?

Both sides of the Great Seal of the United States are shown on the reverse of the one-dollar bill. On the right is an eagle holding both the arrows of war and the olive branch of peace. Above the eagle's head are 13 stars surrounded by a wreath of clouds. The eagle's body is covered with a shield.

On the left is a picture of the reverse of the Great Seal. The pyramid symbolizes permanence and strength. It has not been finished because the United States will always grow, build, and improve. There are 13 layers of stone representing the original states. Each stone within a layer symbolizes local self-government.

The "eye of Providence" above the pyramid is surrounded by light and portrays the spiritual above the material. It also symbolizes freedom of knowledge. Above the eye is a Latin inscription, *Annuit Coeptis,* which means "God has favored our beginnings." The inscription below the pyramid, *Novus Ordo Seclorum,* is Latin for "new cycle of the ages." If you look very closely at the base of the pyramid, you'll see the Roman numerals for 1776.

Neither the schools nor the government has spent much time explaining the meaning of the Great Seal. Hopefully, the next time you look at a dollar bill, you'll have a better appreciation for the Great Seal.

That is, unless the new dollar coin eventually replaces the dollar bill.

Is it true that a man built a castle with stones weighing up to 25 tons with no mechanical equipment or help from anyone? (This is a very weighty subject.)

It's true. One man built an entire stone castle consisting of over 1,100 tons of coral rock, some weighing over 25 tons. He did this amazing feat alone, and in secret.

Edward Leedskalnin was a Latvian immigrant. He was a small man, barely 5 feet tall and weighing around 100 pounds.

When he was 26 years old, his 16-year-old fiancée jilted him on the eve of their wedding. He decided to leave Latvia and emigrate to the United States. He eventually settled in a sparsely populated section of Florida, because he wanted to get away from the world. It is believed that he started building the castle as a way of getting over his lost love, whom he called Sweet Sixteen.

Leedskalnin started building his castle in 1918 and didn't finish it until almost 20 years later. At night he carved huge blocks of coral rock from the land. His only tools were primitive handmade saws, chisels, chains, hoists, hammers, and recycled auto parts. He was fanatical about secrecy and no one ever saw him working. When asked how he could move such huge blocks of stone, he replied that he knew the ancient principles of magnetism used by the Egyptians to build the pyramids.

The castle is made of massive coral blocks fitted together to form a walled, central courtyard. The castle entrance is through a gate, which is a single 9-ton coral block 6.5 feet wide and 7.5 feet tall. When closed it is within a quarter of an inch of the abutting walls. It pivots on an iron rod sitting on an automobile gear. It is balanced so perfectly that a tourist can push it open with one finger.

A tower with a small room on top sits inside the courtyard, which also contains gigantic stone renderings of planets, moons, and suns. One of the most impressive features is a 2.5-ton dining table surrounded by half-ton rocking chairs. The chairs are perfectly balanced, and just touching them lightly will cause them to rock. There isn't a single tool mark on any of the chairs.

To this day, no knows how this small man accomplished such a gargantuan feat. It seems that no one will ever know the secret of the Coral Castle. In 1951 a sign on the castle's front door announced that Leedskalnin had gone to the hospital.

Three days later he died. His secret died with him.

FACTOIDS

The castle contains a 20-foot block of coral shaped like the state of Florida. A small water basin in the table represents Lake Okeechobee. The block is encircled by 10 stone chairs.

When developers planned to build a new subdivision near the castle, Leedskalnin dismantled it and moved it, block by block, 10 miles away. He loaded the huge stones on iron girders mounted on a truck chassis and hired a tractor driver to move the trailer. He never let the driver see him load the stones. One day the driver left at Leedskalnin's instructions, but returned half an hour later. The trailer was empty when he left, but when he came back it was loaded with huge stones.

Through the years many people tried to sneak a look at Leedskalnin while he was working so they could discover his secret. He always seemed to know when someone was watching. No one ever saw how he moved the stones.

Some of the Coral Castle stones are twice as heavy as the largest stone in the Great Pyramid in Egypt.

DID YOU KNOW?

There is another famous castle at the opposite end of the country. It's Scotty's Castle in Death Valley, California.

Walter Scott, known as "Death Valley Scotty," was a prospector who was always looking for gold in the desert. According to him, one day he struck it rich and used some of his new wealth to build a magnificent castle. Although many people tried following Scotty to find the location of his mine, they were never successful. The problem was, there never was a mine.

Scotty happened to meet Albert Johnson, who was in ill health at the time, and the two roamed the desert together. Over time, they became lifelong friends. When Johnson's health improved, he returned East, made a fortune in the stock market, and shared his wealth with Scotty. The castle was in Johnson's name and he used it as a desert retreat.

Johnson never told anyone the truth. He enjoyed the escapades of the flamboyant Scotty, such as the time he rented a train to try to set a speed record.

If you visit Scotty's Castle, make sure you follow the trail behind the main house. It will lead you to the grave of the legendary Death Valley Scotty.

How long was the Pony Express in operation? (Long or not, it embodied the spirit of the West.)

The Pony Express was only in operation a little over 19 months, but it had a profound impact on the country at the time. It not only delivered mail to people in the West but proved that a transcontinental railroad route was possible.

In 1849, gold was discovered in California, and its population soared in just a few short years. Over half a million people lived west of the Rocky Mountains, and they eagerly awaited mail from their families in the east. The government had been searching for a way to develop a transcontinental mail route. They even investigated the possibility of using camels, but that idea proved impractical. Mail took a month or more by boat and almost a month by overland stage from St. Louis, Missouri, to San Francisco, California.

The Pony, as it was called then, was first proposed in 1860 by Senator Gwin of California. The first manager was William Russell of the Russell, Majors, and Waddell Overland Freight Company. He believed that a relay team of riders could carry mail on horseback from St. Joseph, Missouri, to Sacramento, California, in just 10 days. Congress scoffed at the idea and refused to fund it, saying it would be impossible because of snowbound mountain passes and hostile Native Americans.

Russell and his partners pursued the idea on their own. They set up 80 well-provisioned relay stations and carefully selected 400 horses and 80 riders. Because of the fast horses, riders could weigh no more than approximately 120 pounds, saddles and other equipment had to be less than 25 pounds, and no more than 20 pounds of mail could be carried per rider.

On April 3, 1860, the first Pony Express rider placed a message from President Buchanan to the governor of California in his saddlebag and took off on his journey. That saddlebag, with its message, arrived in Sacramento 10 days later. It had traveled almost 2,000 miles.

By October 1861, the telegraph had spanned the nation and the Pony Express was discontinued. In spite of numerous dangers and weather problems, in the entire history of the Pony Express only one mail pouch was ever lost. When the California newspaper, the *Pacific,* paid tribute to the Pony Express, it said, "Goodbye, Pony! You have served us well."

FACTOIDS

Pony Express riders were as young as 11 years old, but none was older than their mid-40s or weighed more than approximately 120 pounds.

Each rider was expected to ride 60 miles in 6 hours. During that time, they used six different horses.

Riders carrying President Lincoln's inaugural address made the fastest run in 7 days and 17 hours.

In spite of its fame, the Pony Express was a financial disaster. The founders had invested $700,000 and ended up with a debt of $200,000. It was sold at an auction to Ben Holladay, who four years later sold it to Wells Fargo for $2 million.

The Pony Express designed its own saddle and mail pouch, called a *mochila.* None is known to exist today.

The legendary Buffalo Bill was a Pony Express rider.

Morgan horses and Thoroughbreds were used at the eastern end of the trail, pintos in the middle section, and mustangs on the western end. Most horses were mares.

To jump off his horse, transfer his mail pouch, and leap on a fresh relay horse took a rider two minutes or less.

DID YOU KNOW?

Pony Express riders faced many dangers. Not only did a rider frequently travel 100 miles a day, he often fought raging blizzards, had trouble finding the trail because of snow, risked death riding near the edges of steep canyons, and faced possible ambush from hostile Native Americans.

There are many tales of bravery and stamina. One such tale is that of Bob Haslam. One day Haslam received the eastbound mail from San Francisco and galloped off. When he reached the relief station, the next rider was so frightened of hostile tribes that he refused to ride. Haslam kept riding to the next station. He rode 190 miles without stopping to rest.

After resting for nine hours, he started on the return trip. The next station had been raided, the station keeper was dead, and the horses were gone. Haslam kept riding until he reached his original starting point. He had made a 380-mile round trip, the longest on record.

Haslam was ambushed on another trip. An arrow pierced his arm and another plunged into his face, fracturing his jaw and knocking out his teeth. He refused help at the next station and continued on until he reached his final destination. Although badly wounded, he had ridden 120 miles in a little over 8 hours.

On one trip, Haslam's way was blocked by a band of 30

Paiute Indians. The chief knew Haslam's reputation for bravery and let him pass, saying, "You pretty good fella—you go ahead."

Bob Haslam and riders like him embodied the Pony Express slogan, "The mail must go through!"

More questions? Try these Web sites.

PONY EXPRESS
http://www.xphomestation.com/

A wonderful site with a thorough discussion of the Pony Express. It covers the history, the riders, horses, station keepers, salaries, weapons, and the mail. It has interesting stories about the Paiute War, the longest ride, and the fastest time. It also has short biographies of many of the Pony Express riders.

U.S. PRESIDENTS AT A GLANCE
http://homepage.midusa.net/~rlong/pres.html

This site lists all 42 presidents of the United States. Just click on any president's name for a short biography and the highlights of his presidency.

CORAL CASTLE
http://www.parascope.com/en/articles/coralCastle.htm

This site has an excellent article about the Coral Castle in Florida. It includes a number of color photographs.

TOURIST ATTRACTIONS IN THE UNITED STATES
http://www.go-unitedstates.com/

This handy site divides the country into seven regions and lists the popular attractions for each region. For example, the Southwest section includes the Grand Canyon, Disneyland, Dinosaur National Monument, Hoover Dam, the USS Arizona Memorial, the Hansen Planetarium, and White Sands national monument. Clicking on any attraction takes you to a Web site devoted to the attraction. There is also a link to the National Park Service.

The left side of the page includes museums, events, galleries, theaters, and night clubs. Clicking on any one subject will bring up a new page, again divided into regions of the country. For instance, if you click on museums and scroll down to the Southwest, you see links to various museums in Arizona, California, Colorado, Hawaii, Nevada, New Mexico, and Utah.

15

Weather

What are the differences among sleet, freezing rain, and hail? (Don't forget about the graupels.)

Rain normally falls as drops of water. However, if the rain falls through air below 32°F, it freezes. These frozen drops of water are known as sleet.

Freezing rain, on the other hand, forms near the ground. If the rain happens to fall through a thin layer of cold air near the surface of the earth, the raindrops cool to temperatures below freezing. In this case they do not freeze and turn into ice, but remain liquid. This phenomenon is called supercooling. When this supercooled rain strikes power lines, trees, buildings, the ground, or any exposed object, it freezes instantly, covering everything with a thin layer of ice.

Hail usually forms during thunderstorms. Raindrops are blown high up into the cold areas of the clouds where they freeze. As they fall back down, more water adheres to them. When they are blown back up into the cold cloud again, the layer of water freezes, adding another coating of ice. This process keeps repeating itself until the hail is so heavy that the wind currents can no longer support the weight and the hail falls to earth.

There is another object that people usually mistake for hail. It's called a graupel. When a snowflake is falling to earth, supercooled drops of rain may sometimes freeze to its surface. These frozen snowflakes are called rime. If more frozen drops of water accumulate on the snowflake, it forms a mass known as a graupel. A graupel is like a miniature snowball, unlike hail, which is a solid piece of ice.

FACTOIDS

In 1888 a hailstorm killed over 250 people in the town of Moradabad, India.

The heaviest hailstones ever recorded fell on Bangladesh in 1986. They weighed up to 2.25 pounds and killed 92 people.

In 1970 Coffeyville, Kansas, was hit by the heaviest hailstone ever known to fall in this country. It weighed 1.5 pounds and was the size of a grapefruit.

When hail falls, it can reach speeds up to 100 mph.

If you cut a hailstone in half, you'll see concentric rings similar to those of an onion. The rings indicate how many times the stone traveled to the top of the storm before falling to earth.

Every year hailstorms cause around one billion dollars in damage to crops and property in the United States.

One of the worst hailstorms in the history of the country hit southeastern Iowa in 1925. The hail destroyed crops in an area almost 10 miles wide and 75 miles long and killed poultry and livestock. Damage was estimated at $2.5 million. Cornfields were so flattened that many farmers had to leave their farms to find work elsewhere.

DID YOU KNOW?

An ice storm can transform the landscape into a fairyland as trees are draped with glistening white lace and sparkling threads of ice hang from telephone wires. An ice storm can also be devastating, causing enormous damage to life and property.

When it comes to ice storms, 1998 produced what became known as the ice storm of the century and was in fact the worst ice storm ever recorded.

Actually, it wasn't a single storm, it was a series of storms that hit one after another. There was no time for the freezing rain to melt between storms. The storms continued for almost a week. They hit northeastern Canada and New England with a severity rarely seen in that area.

The storms coated everything with one to three inches of ice. The quiet winter was shattered by the sound of ice-coated tree limbs crashing to the ground, taking out power lines in the process, and plunging major cities into darkness. Under the weight of tons of ice, entire trees fell and smashed into cars and houses. Road travel was virtually impossible because of the ice and debris covering streets and highways.

With one out of four homes without power, temperatures dropped, sump pumps failed and basements flooded, food thawed out in freezers, water pipes froze and burst, and many residents were forced to seek refuge in shelters. In Canada alone there were over one million homes without power. In the city of Quebec, some homes were without power for a month.

When the ice storms were over, damage was estimated at

over one billion dollars. Canada and the Northeastern United States are the largest producers of maple syrup in the world, but many orchards were leveled. If new trees are planted, it will take 25 years before they can produce maple syrup again.

Ironically, the Ontario Hockey League had to cancel their games. The ice in the skating rink had melted because there was no power to run the refrigerating unit.

Did a rainmaker ever actually cause a flood in San Diego? (The city council thought he was all wet.)

In 1912 San Diego was experiencing a very dry year. The Morena Dam reservoir was less than one-third full, the city's growth depended on an abundant water supply, and another drought was imminent. A real estate agent had heard of Charles M. Hatfield's reputation for making rain and suggested that the San Diego City Council enlist his aid.

Hatfield called himself "the Moisture Accelerator," but to those who were familiar with his work he was simply "the Rainmaker."

In the early part of the 20th century, Hatfield had traveled from California to Alaska and was considered the most successful rainmaker of the time. He had been successful in bringing rain to the parched farmland of the Central and San Joaquin valleys of California. In 1904 the Los Angeles Chamber of Commerce said they would give him $50 if he could cause it to rain in that city. He set up his apparatus and promised that the rain would fall in less than five days. On the fourth day over an inch of rain fell on the dried-out city.

After arguing for almost three years, the city of San Diego decided to enlist Hatfield's help. He promised to produce 40 inches of rain provided the city paid him $10,000. They agreed.

It was January 1915 when Hatfield and his brother started

mixing the secret chemicals at a site 60 miles east of the city. The next morning, rain clouds had formed and by noon there was a downpour, filling neighboring ponds and streams. Hatfield continued his work. Bridges washed away and homes were flooded. A stranded Santa Fe train had to unload passengers by boat. And the rain kept coming down.

A city official decided that enough was enough and tried to telephone Hatfield to tell him to stop, but the telephone lines were down because of the flooding.

By the fourth day the land was saturated. More bridges washed away, railroad tracks loosened, all roads within a 60-mile radius were submerged, and muddy water covered farm and ranch lands. Finally, the rain stopped, but only for a few days.

When the rain started again, one dam simply disappeared, another dam ruptured, and the water at Morena Dam starting rising at the rate of two feet per hour. Frantic dam engineers opened the two main safety valves to let water run out of the dam into the canyon below. When the rain stopped for good, the water in the dam was only five inches from the top. Some estimated that enough water had been released from the dam to fill it twice.

Surveying the damage and hearing that a lynch mob was after them, Hatfield and his brother used aliases when entering the town. The San Diego City Council refused to pay his fee, fearing that if they did they would be responsible for the millions of dollars in damages caused by the flood. Hatfield countered that he had delivered what he had promised and the city should have taken more precautions.

Hatfield lost the court cases that followed. The California Supreme Court ruled that the rain was "an act of God" and not an act of Hatfield's. He never received a penny for producing one of the greatest rainstorms in San Diego history.

When Hatfield died in 1958, his rainmaking secret died

with him. Today a small granite plaque with the words HATFIELD THE RAINMAKER stands near the Morena Dam.

FACTOIDS

In 1994 during Super Jam IV at Athens, Georgia, the band Widespread Panic was ready to play the song "Hatfield" (based on the story of the famous rainmaker). As they started playing the sky got dark, and before the song ended it was raining so hard that the generators had to be shut down and the band was forced to leave the stage. Perhaps Charles Hatfield was plying his trade again.

The official name for a rainmaker is "pluviculturalist."

Some think that Burt Lancaster's role in the 1956 film *The Rainmaker* resembled Hatfield.

DID YOU KNOW?

In the history of almost every civilization there is the story of a great flood. Many aspects of the story are similar. People have become evil and must be destroyed. The few good people are saved by building a boat, climbing a mountain, or climbing a tree.

One such flood story is from an ancient Mesopotamian myth. Enlil, the senior diety, is upset with humans because they make so much noise, so he decides to wipe out the entire human race. He tries various plagues, but none of them does the job. Finally, he decides on a flood. Enki, the god of wisdom, warns Atrahasis, the king. Atrahasis, his family, his possessions, and animals and birds all board a reed boat. They ride out the flood that lasts for seven days and seven nights. When the waters subside, Atrahasis makes offerings to the gods. Enlil finally gives in and decides to let the human race continue, provided Enki and his mother make sure the humans no longer bother Enlil with their noise.

What is ball lightning?
(Don't try playing catch with it.)

We typically think of lightning as a "bolt" that streaks down from the sky to the earth in a brilliant flash. However, sometimes a blazing ball will move through the air parallel to the earth. This luminous ball, which can be as small as an orange or as large as a cantaloupe, is called ball lightning.

Although it usually moves parallel to the earth, ball lightning will often make vertical jumps. It not only descends from thunder clouds but can often magically materialize indoors or outdoors.

Until recently, the scientific community did not believe that ball lightning existed because of its strange behavior.

The ball moves rather slowly, a few miles per hour, floats about three feet off the ground, and can make erratic changes in direction. It is not affected by the wind and may move into the prevailing wind rather than away from it. It may last for less than a second or as long as several minutes, but it typically has an average life of around 25 seconds. It has been seen in houses, in airplanes, and passing through closed glass windows without affecting the glass in any way. If it is blue or orange, it tends to last longer than if it is some other color. The ball either disappears silently or extinguishes with a violent explosion that often causes extensive damage.

Sightings of ball lightning have been well documented and have been reported since the time of the ancient Greeks. People who have seen this strange phenomenon claim that it is truly beautiful. Witnesses also note that there is a bad smell at the same time.

For decades, scientists have proposed theories about the nature of ball lightning. However, there is still no accepted theory. The origin and nature of ball lightning remain a mystery even today.

FACTOIDS

There are normally 2,000 thunderstorms in the world at any point in time, producing over 100 flashes per second.

Each year, approximately 10,000 forest fires are started by lightning.

The largest number of people killed by lightning at one time was in 1963 when lightning hit a Boeing 707 jet. It crashed with the loss of 81 passengers.

Of all people killed by lightning, 84 percent are males.

There are twice as many lightning casualties in Florida as any other state.

Although most doctors know how to treat victims of electrical shock, very few are familiar with keraunopathy, or treatment of lightning strike victims.

Contrary to popular opinion, rubber-soled shoes and rubber automobile tires offer no protection against lightning. However, the steel frame of an automobile will provide some protection if you are not touching any metal. You are typically safer inside your car than outside it.

DID YOU KNOW?

If you get hit by lightning and survive, your troubles may not be over. First, you may have to be treated for burns, contusions, and fractures. Injuries like these are easy to recognize and treat. However, problems that are not apparent at first can later cause you a great deal of suffering.

Some of the effects of being hit by lightning include short-

or long-term memory loss, disturbances in your sleep pattern, depression, fatigue, thought impairment, muscular pains, severe migraine or cluster headaches, loss of dexterity, hearing impairment, seizures, and irritability.

It's easy to see why getting struck by lightning would make someone irritable. After all, there's enough electricity in a lightning bolt to power 10,000 electric chairs.

If getting hit just once would make someone irritable, you can just imagine how irritable a former U.S. park ranger, Roy C. Sullivan, became. He was struck by lightning on seven different occasions. It all started in 1942 when his only injury was the loss of his big toenail. Over the next 35 years, he was struck by lightning six more times with various injuries, including a burned shoulder, hair set on fire (twice), leg burns, chest and stomach burns, and a wounded ankle.

But lightning never did manage to kill him. He committed suicide in 1983 because he was allegedly rejected in love.

What is the heat index? (This is a hot topic.)

The temperature alone does not always tell you how cold or hot it is. The heat index is to summer what the wind chill factor is to winter.

During cold weather, if wind blows across your skin, it strips the insulating layers of warm air and replaces them with colder air. The faster the wind, the colder you feel. For instance, if the actual temperature is −15°F and there is a 20-mph wind, then it will feel as if the temperature is −60°F.

During hot weather, heat and humidity can combine to make you feel much hotter than the thermometer indicates. When you perspire, the perspiration evaporates and gives a cooling effect to your skin. If the humidity is too high, perspiration cannot evaporate and you feel hotter. Sometimes during hot weather, water from the air condenses on your skin and makes you feel even hotter still. For instance, if the actual temperature is 95°F and the relative humidity is 80 percent, then the heat index, or the heat you will feel, is 115°F.

Temperatures from 100° to 110°F can be hazardous, and temperatures above 110°F can be downright dangerous. In fact, if you are doing physical activity or are exposed to heat for a long period of time, you can suffer from sunstroke, heat cramps, or heat exhaustion whenever the heat index rises above 90°F.

Here are the heat index temperatures that can cause you problems.

Risk	**Heat Index**	**Danger**
Be careful	80–90	You'll get tired faster when you exercise.
Caution	90–105	You can get either heat cramps or heat exhaustion.
Danger	105–130	It's very likely you'll get heat exhaustion.
Extreme danger	130 or more	You're definitely going to get heat stroke.

To avoid these dangerous conditions, you should always rely on the heat index rather then the temperature shown on the thermometer. When the heat index is high, drink a lot of water and juice, wear light-colored and loose-fitting clothing, stay out of the sun and in an air-conditioned area if possible. If you do these things, you won't risk maladies caused by the heat and you'll feel a lot better besides.

FACTOIDS

Extreme heat can kill. Every year about 175 people in the United States die from heat-related conditions. During the summer of 1901, there were over 9,500 heat-related deaths across the Midwest, and in the summer of 1988, an estimated 10,000 people in the East and Midwest died from the heat.

Men are more susceptible to illnesses caused from heat because they perspire more and dehydrate more quickly.

The highest recorded temperature in the world was 136°F at Al Azizia, Libya, on September 13, 1922. The highest recorded temperature in the United States was 134°F at Death Valley, California, on July 10, 1913.

DID YOU KNOW?

Most people don't know the differences among heat cramps, heat exhaustion, and heat stroke. The following information might be helpful.

Heat cramps are usually caused by exercising in hot weather. You will perspire profusely and have painful cramps or spasms in your legs or abdomen because your body salts become unbalanced. As you become used to the heat, the cramps will not occur as often. The best cure is to just take it easy.

Heat syncope is just a fancy term for fainting. If you're not used to exercising and it's very hot, the heat can cause a fast drop in blood pressure and you will pass out. The cure is the same as for heat cramps; take it easy.

Heat exhaustion is due to your losing fluid and salt because of excessive perspiring or because you aren't drinking enough fluids. If you suffer from heat exhaustion, you will perspire profusely and become very weak, and your skin will be cold, clammy, and pale. You will probably vomit and might even faint. Take it easy and drink plenty of water.

Heat stroke (sunstroke) can kill you! The extreme heat prevents your body's thermostat from working properly. Your skin will be dry and hot, and your temperature will be 106°F or higher. You'll become confused and lethargic, and you'll probably lose consciousness. Seek immediate medical aid. Even if you merely suspect a heat stroke, don't take a chance. Get medical help right away.

When temperatures drop below freezing and blizzards howl in the night, most people try to stay indoors and keep warm. If they have to venture outside, they bundle up and take every precaution they can against the cold.

However, people react much differently in the summer. They enjoy hot weather and like basking in the sun. They play sports and exercise. They seem to think that no matter how hot it gets, the worst thing that can happen to them is a bad sunburn. That's not true. Excessive heat can make you sick and even kill you.

Some simple advice will keep you from getting heat-related illnesses: stay cool!

What was the "year without a summer"? (Sunsets and Frankenstein.)

The year 1816 was known as the "year without a summer." There was no summer at all in New England that year. There was snow and ice in April, ice half an inch thick on lakes and ponds in May, snow up to 10 inches deep in June, and more ice in July and August. The frost and ice continued through September, October, and November. The weather didn't become mild until December, but by then it was too late. Although a few warm summer days lulled farmers into planting crops, frost came a few days later and killed the plants before they could grow into maturity. Because most crops were destroyed, people later called the year "eighteen hundred and starve to death."

In the United States, New England was the hardest hit. Because of the lack of feed, livestock died and farmers were forced to eat what they could find, including pigeons and mackerel purchased from coastal fishermen. What grain was available cost 10 times as much as usual. Crops were also small in the Midwest and farmers had to pay up to $5 a bushel for seed for planting in the spring.

The severe winter wasn't limited to this country. It circled the globe and affected Western Europe and the Scandinavian countries. Famine was widespread and riots broke out everywhere. Looters emptied grain warehouses and many people searched the streets for stray animals to eat.

Many blamed Ben Franklin for the severe weather. They claimed that his experiments with lightning rods had funneled off heat from the sun. However, the culprit was not Ben Franklin, but Tambora.

Mount Tambora, a volcano on Sumbawa Island in Indonesia, erupted in April 1815. It was the largest volcanic eruption since the beginning of recorded history. The explosion was heard almost 1,000 miles away, smoke could be seen from a distance of over 300 miles, and volcanic ash fell on areas as far away as 800 miles.

Falling rock and poisonous gas killed 10,000 people on the island of Sumbawa. In the ensuing weeks, an estimated 80,000 people on neighboring islands died from starvation due to crop damage from the volcanic ash.

The volcano threw over a million and a half tons of dust and debris into the sky, which spread over a million square miles. The clouds of dust blocked out light and heat from the sun and caused the year without a summer.

The only good thing about the eruption of Mount Tambora was that it created spectacularly brilliant and beautiful sunsets throughout the world for several years.

FACTOIDS

When Mount Tambora erupted, it reduced the mountain from over 13,000 feet to 9,255 feet and created the world's largest volcanic cone.

Today, Sumbawa Island is a tourist attraction and hosts such events as buffalo races over newly planted rice fields. The eastern capital of Bima is home to strong ponies exported all over Indonesia to pull carts. Bima also exhibits *ntubu,* a traditional game of head fighting, in which opponents pound their heads against each other. It is so dangerous that a priest performs an immunity ceremony before the game.

Just to the east of Sumbawa is Komodo Island, famous for its dreadful Komodo dragons. Because of the reputation of these monitor lizards, there are few visitors to the island and even fewer inhabitants.

Bungin Island is not too far from Sumbawa and is known as the world's most densely populated island. About two thousand five hundred people live on 37 acres. When someone wants to build a house, they simply import coral rubble and pile it on the edge of the island as a foundation to build on.

DID YOU KNOW?

The year without a summer produced more than cold weather and beautiful sunsets. Mary Shelley spent the summer of 1816 in Switzerland. She and her poet husband, Percy Bysse Shelley, had been visiting Lord Byron and some friends at the Villa Diodati. Because of the eruption of Mount Tambora, the tempestuous weather produced torrential rains and incredible lightning storms that June night. The Shelleys decided it was too stormy to risk traveling home so they spent the night.

For entertainment the group read ghost stories to one

another. This gave Byron an idea and he challenged everyone to write a ghost story of their own. This went on for a few nights but Mary Shelley didn't feel inspired and wrote nothing. Six days later, when the others had abandoned their stories, Mary Shelley had an inspiration and wrote, "It was on a dreary night of November."

She completed her novel a year later. It was called *Frankenstein or the Modern Prometheus.*

Why is a rainbow curved? (Straight up and down would look silly.)

A rainbow is created when sunlight is refracted through drops of water. In other words, the water bends the sunlight, and each color frequency bends at a slightly different angle. The water acts like a prism, breaking the sunlight up into different colors. The colors are always the same. Starting from the outside, or top, of the rainbow, the colors are always red, orange, green, blue, indigo, and violet. The larger the drops of water, the brighter the rainbow will be.

A rainbow is curved because of the way light is refracted in water. For example, consider a single, round drop of water. As light enters the drop, it is refracted, or bent, toward the center of the drop. When it hits the inside wall of the drop, it is again refracted and starts a return path, but it hits the other side of the drop and is once again bent as it leaves the drop. To understand how this works, imagine a large glass bowl standing on a table. If you dropped a small ball into one side of the bowl with enough force, it would roll down toward the center of the bowl and then up the other side in a semicircular path. The ball follows the curvature of the bowl, which is a half circle. That's just what happens when light hits a drop of water—it follows the curvature of the drop and forms a half circle, or bow.

It really doesn't matter what causes a rainbow to be curved. It is always a joy to see such a wonderful and colorful creation of nature.

FACTOIDS

Because a rainbow is a distribution of colors with reference to the eye of the person looking at it, no two people ever see the same rainbow (except in a photograph). In fact, each eye actually sees a different rainbow.

In ancient times, many people were afraid of rainbows. They thought a rainbow was a group of snakes rising in the sky to drink water. They believed that if you pointed at a rainbow, you would lose your finger.

Although extremely rare, it's possible for a rainbow to appear at night if there is a full moon and rain. Because moonlight is much weaker than sunlight, the rainbow will be very faint.

In some cultures in Indonesia and Africa, a rainbow is believed to be a heavenly bridge that connects the world of humans with the world of the gods. To many Arabs, it is the bow of God, and to the Masai it is the robe of God.

An old German creation myth says that the rainbow is the bowl that holds the paint God uses when coloring the birds.

Because rainbows are so distant, they are hard to observe closely. To solve that problem, scientists in the Middle Ages filled large glass globes with water and let light shine through them so they could study rainbows.

DID YOU KNOW?

You've probably heard the expression "the pot of gold at the end of the rainbow." The question is "Who put the pot of gold there?"

According to Irish folklore, the pot of gold belongs to a leprechaun, an Irish fairy. The typical leprechaun is only about two feet tall and looks like an old man. He tends to be very unfriendly, even nasty, and lives alone. He spends most of his time making shoes and wears a leather apron. It's said that he has a hidden pot of gold.

If you want to find the pot of gold and you don't want to chase rainbows, you will have to find a leprechaun. If you listen carefully, you'll hear the tapping of his hammer while he's making shoes. Should you manage to catch a leprechaun, hold on tight and don't let him get away. If you threaten to hurt him, he'll probably tell you where his gold is hidden. However, never, never, take your eyes off him, because if you do he will vanish. The leprechaun will use every trick he can to make you look away so he can escape.

Whether it's capturing a leprechaun or looking for the pot of gold at the end of a rainbow, finding treasure is not easy. Even if there were a pot of gold at the end of a rainbow you'd never be able to get it. As you walk toward the rainbow, it will always move away from you.

People who are seeking easy riches, or dreamers, are often accused of "chasing rainbows." Perhaps the way to wealth requires hard work and people should be more practical about it.

However, the dreamers in our society have created wonderful art, buildings, and inventions. It might not be a bad idea after all to search for that wonderful land somewhere over the rainbow.

More questions? Try these Web sites.

Ball Lightning

http://www.eskimo.com/~billb/tesla/ballgtn.html

This has a great deal of information about ball lightning, including research reports and personal encounters. There is also a section on how to create your own ball lightning, and a list of other Web sites about ball lightning.

Rainbow

http://www.unidata.ucar.edu/staff/blynds/rnbw.html

A very informative site that describes what a rainbow is, what makes it curve, what makes the different colors, what causes a double rainbow, and more.

Heat Index Calculator

http://www.Web100.com/~sib/heatindex.html

If you enter the current temperature and the relative humidity, this calculator automatically computes and displays the heat index.

Tidal Waves

http://tqjunior.advanced.org/5818/tidalwaves.html

Did you know that a tidal wave can travel as fast as 200 mph and can be up to 1,000 feet high? These and other interesting facts about tidal waves are covered in this site.

If you get bored with tidal waves, you can also find information about avalanches, tornadoes, hurricanes, floods, thunderstorms, and other weather-related topics.

Your Weather

http://www.weatherforyou.com/

This is an excellent weather site. It has national weather

information, and you can also enter your city and state for a local weather forecast. It includes a weather computer that lets you convert Celsius to Fahrenheit and vice versa. This site also has calculators for determining the wind chill and the heat index.

16

World

Is it true that the only man-made object on earth that you can see from outer space is the Great Wall of China? (It's amazing what you can see from outer space.)

There is a popular misconception that you can see the Great Wall of China with the naked eye if you're in outer space. It's not true. When astronaut William Pouge orbited 300 miles above the earth in the Skylab space station, he could not see the Great Wall of China without binoculars. Space shuttle astronaut Jay Apt said that his team looked for the Great Wall of China but could not find it, even though they could see things as small as airport runways. He deduced that the wall's material was the same color as the ground and blended into the surrounding soil, making it impossible to see. At the time, the space shuttle was only 180 miles above the earth.

Many man-made objects on earth can be seen from outer space with the naked eye. Astronauts have been able to see the rocket-sled test site in New Mexico, the aircraft carrier that was scheduled to pick them up, and highways, railroads, and canals.

The building of the Great Wall was one of the world's largest construction projects. Construction started in the 7th century B.C., when various feudal states each built their own walls for defense. When China was unified in 221 B.C., all of the existing walls were joined to ward off invaders from the North. After that, the wall was renovated many times. When the Ming Dynasty was founded in 1368, a major renovation of the wall was begun. It took 200 years to finish the work and resulted in the wall that can be seen today. The wall was used to stave off nomadic invaders and to communicate with the capital. Messages (smoke by day, fire by night) were relayed from tower to tower.

FACTOIDS

The wall is an average of 25 feet high, 15 to 30 feet wide at its base, and about 12 feet wide at the top. Its watchtowers are about 40 feet high.

An imaging radar on the Endeavor Space Shuttle produced images that not only clearly showed the Great Wall but also showed remnants of an older version of the wall.

According to ancient records, at least one million people worked on building the Great Wall. Many laborers died from starvation and exhaustion. Rather than being buried properly, their bodies were added to the wall's building material to save time. Today, archeologists have discovered tombs within the

Great Wall, which has often been called "the longest cemetery in the world."

The wall is 3,976 miles long, as long as a wall built to extend from San Francisco, California, to Caracas, Venezuela.

Because the Great Wall is actually a series of different walls, different sources give different lengths for the wall. Most authorities agree that it is 3,976 miles long. However, Richard Nixon said that it was 2,484 miles long, and *Time* magazine once reported that it was 1,684 miles long.

DID YOU KNOW?

Evidence of the Romans in northern Europe is also a wall. The Roman emperor Hadrian ordered a wall to be built in northern England to protect Rome's possessions from marauding tribes. Built by Roman troops, it was 73 miles long and was called Hadrian's Wall.

Unlike the Great Wall, Hadrian's Wall was built entirely from stone. It took over 15 years to build the wall, which was 8 to 10 feet thick and 12 to 16 feet high. Towers were located approximately one-third of a mile apart with small forts every mile. In spite of the towers and forts, the Romans never intended to fight from the top of the wall. Soldiers were trained to meet the invaders on open ground. The wall was primarily intended to control the movement of people across the frontier.

The main invaders were the Picts, who always seemed to be at war with the Romans. Because they had a custom of painting their bodies, the Romans used the Latin word *picti,* which means "painted," to describe them. The Picts breached the wall three times and it was finally abandoned.

Many parts of Hadrian's Wall can still be seen today and are quite impressive. Although the wall does not separate Scotland from England, many local Englishmen refer to Scotland as "the other side of Hadrian's Wall."

There have been many famous walls, including the Great Wall of China, Hadrian's Wall, the Berlin Wall, and others. They were all designed to keep people out, or keep them in.

It's a good thing that we aren't building great walls today.

What are the seven seas? (A lucky number for sailors?)

An ocean is a large area of salt water unobstructed by continents, such as the Pacific Ocean and the Atlantic Ocean. A sea, on the other hand, is usually partially or completely enclosed by land, such as the Dead Sea and the Mediterranean Sea.

Sometimes oceans are referred to as seas. When someone talks about the "seven seas," they are referring to the seven oceans, which are: the Arctic, Indian, North Pacific, South Pacific, North Atlantic, South Atlantic, and the Southern oceans, the last lying below Australia and the southern end of South America.

Years ago, when sailors had sailed all of the seven seas, they were considered to have sailed around the world.

FACTOIDS

Each of the seven seas, or oceans, is different from the others and presents unique challenges to sailors.

Arctic Ocean: This sea is the smallest and is nearly landlocked. For most of the year it is covered with ice, forcing sailors to face subfreezing temperatures. There is little marine life beneath the ice, although it is abundant in areas of open waters. This sea is the shortest route from Russia to North America.

Indian Ocean: This is the third largest after the Atlantic and Pacific oceans. It includes both the Red Sea and the Persian Gulf. Because it is an important trade route between Asia and Africa, it has been the scene of many conflicts. Sailors often must contend with monsoons and occasional typhoons.

North Pacific Ocean: This sea contains the lowest point anywhere on earth, the Mariana Trench, over 30,000 feet below sea level. Sailors often face persistent and hazardous fog for seven months out of the year.

South Pacific Ocean: When sailing this ocean, sailors may encounter dangerous and often deadly typhoons and must dodge icebergs when sailing the most southerly routes. Most of the world's islands, about 25,000 of them, are in this ocean.

North Atlantic Ocean: This ocean has large areas of continental shelves and is relatively shallow compared to other oceans. It is known for its violent and often fatal storms. Sailors must fight icing of their ship's superstructure for eight months, from October through May.

South Atlantic Ocean: When sailing this ocean, sailors must watch for hurricanes. If they sail too far south, they will encounter icebergs and icing of their ships.

Southern Ocean: This is the only ocean that encircles the entire earth without being blocked by some land mass. Sailors have to contend with huge icebergs, some of which tower 10 stories above the water. They must also face 12-foot swells and waves often as high as a 7-story building. Fierce gales, often reaching speeds of over 70 mph, occur 20 percent of the time. It is not a hospitable ocean at all.

DID YOU KNOW?

Long ago, people managed to divide the world's water surface into seven oceans. Why seven? Perhaps it's because many cultures find seven to be a mystical number.

Some authorities believe that the number seven indicates spiritual perfection. They claim that the Hebrew word *shevah,* meaning "seven," is derived from the root *savah,* "to be full" or "satisfied."

The Bible contains 424 references to the number seven, from Genesis to Revelations. Most of us know many of them, such as God resting on the seventh day after creating the world. Here are some lesser-known examples:

There were seven years of plenty throughout Egypt followed by seven years of famine.

The priest of Midian had seven daughters.

For seven days you shall eat unleavened bread.

And you shall offer with the bread, seven lambs.

I will punish you seven times for your sins.

Build me here seven altars, and prepare me seven oxen and seven rams.

The Bible is not the only place where the number seven is found. It occurs everywhere. For instance, we have the seven hills of Rome, the seven seas, the seven dwarves, the seven deadly sins, the seven wonders of the ancient world, and seven days in a week. Any gambler knows that rolling a seven with the dice is lucky. There are many, many more examples.

If you ever decide to sail the ocean and encounter frightening weather, remember that you are sailing on one of the seven seas, and seven is a very lucky number.

Who was Hannibal and why was he crossing the Alps with elephants? (It's better than walking.)

Carthage and Rome were bitter enemies and had been engaged in a series of wars for about 100 years. Hamilcar was the son of a great Carthaginian general, Hamilcar Barca. When Hannibal was nine years old, his father made him swear eternal hatred for the Roman Empire.

Hannibal eventually became a great general and spent his entire life in a constant struggle against Rome. He knew that war with Rome was inevitable, but he determined to fight at a time that was advantageous to Carthage and to fight the war on his own terms.

The fact that Hannibal and his army were in Spain did not worry the Romans. They thought that no one could possibly invade their country by land. To do so, an army would first have to fight its way through a Roman army and then cross the imposing Pyrenees mountains. If they managed to do that, they would then have to fight across the south of France, controlled by the Romans, to reach the final barrier, the formidable Alps. All in all, a virtually impossible task.

Hannibal had approximately 40,000 troops and 40 elephants. He managed to elude the Roman army in Spain and approached the Rhône River. The Romans still did not panic. They sent a second army to secure the bridges at the Rhône. Once again Hannibal fooled them. He sneaked northward, avoided the Roman sentries, and crossed the river on pontoons and by swimming. It was spring and the river was flooding and very treacherous. Many of the elephants drowned. Nevertheless, Hannibal and his army crossed the river, marched back south, caught the Roman army by surprise, and was victorious. His last obstacle now faced him. The Alps.

Crossing the Alps was a remarkable achievement. Not only were the mountains themselves dangerous but local tribes fought anyone who ventured onto their lands. Plagued by snowstorms and landslides, Hannibal had to fight his way through the mountains. When he arrived in Italy, he had only 26,000 troops left and about two dozen elephants. The other men and elephants had either drowned in the Rhône River or had been killed by marauding tribes.

By now, the Romans were frightened. They sent troops to

meet Hannibal, who was now outnumbered two to one. Yet with great cunning he defeated the superior force. He outfoxed the Roman forces at every turn. Finally, the Romans maneuvered him into a battle in an open field where Hannibal could not surprise them. It didn't work. Hannibal used a classic strategy to defeat the Romans once more. He first let the Romans penetrate the center of his infantry, then surrounded the Romans with his cavalry and attacked. Of the 70,000 Roman soldiers, only 10,000 survived the battle.

It took another 12 years before the Romans finally defeated Hannibal. He went down in history as a legendary commander who performed one of the greatest feats in military history. The only picture of him that survives to this day is an image on a Carthaginian coin. It is the picture of a young man with a pleasant face.

FACTOIDS

The Alps contain many sites where ancient peoples have engraved pictures into the rocks. At one site, it is estimated there are between 200,000 and 300,000 rock carvings. At another site, scientists have found more than 35,000 of these art carvings.

The Alps are a playground for millions of visitors in both summer and winter. The impact of so many human visitors has severely degraded this fragile environment. The Alps are now the most threatened mountain system in the world.

Although elephants can swim, the torrents of the Rhône River would have swept them away and they would have been lost. Hannibal filled bladders with air and put them on the elephants to float them across the river on their "water wings."

DID YOU KNOW?

One of the most famous mountains in the Alps, and a favorite of mountain climbers, is the Matterhorn, the fourth highest mountain in Europe.

If you look at the mountain from the Swiss side, it appears to be a horn-shaped peak. If you look at it from the Italian side, it resembles a classic pyramid. If you don't get a chance to visit Europe, there is a faithful replica of the Matterhorn at Disneyland. Although it is a hundredth the size of the real mountain, it is still an impressive 147 feet high, and the use of forced perspective makes it look even higher.

In 1840 a British mountaineer, Edward Whymper, was the first person to climb the Matterhorn. In a subsequent climb four members of his party were killed when one of them slipped and pulled three others to their deaths. Fortunately the rope broke before it pulled down Whymper and his two guides. This is still one of the best known mountaineering accidents.

In spite of his name, Whymper was certainly no wimp.

What is the story of Rome, Romulus, and Remus? (Part of a wolf pack?)

According to Roman mythology, Romulus and his twin brother, Remus, were the founders of Rome. In the legend, two brothers, Amulius and Numitor (the twins' grandfather), were heirs to a kingdom. Amulius took the kingdom away from his brother, but because he was afraid that Numitor's heirs might one day dethrone him, Amulius had all of Numitor's male children killed. He spared the life of Numitor's daughter but consecrated her to the temple of Vesta so she would never marry and have children. The god Mars was upset with Amulius's cruelty and gave Numitor's daughter two sons, Romulus and Remus. Hearing about the birth of twins, Amulius was infuriated and ordered the babies to be thrown into the river and drowned.

A servant carried the babies in a basket to the river, where he was frightened by the raging water, and he set the basket on the river's edge without throwing the twins into the water. The flooding river swept up the basket and carried it to a shallow pool where it became stuck in the roots of a fig tree. A female wolf heard the babies crying, licked the mud off them, and, with the help of a woodpecker, looked after them until they were eventually found by a shepherd who raised them as his own children.

When Romulus and Remus grew older, they discovered who they were and recruited a band of youths to kill Amulius and restore their grandfather Numitor to the throne. After successfully accomplishing their mission, they decided to build a town on a hill near the fig tree that had saved them. The two brothers waited for a divine sign over the hill to decide which of them should build the town. Remus saw 6 vultures, but Romulus later saw 12 vultures, so he was made king and started building the city.

After the town was built, Remus became furious because he hadn't been made king. He fought with his brother and was killed. With Remus dead, Romulus decided to name the town Rome, after himself.

Romulus ruled for many years until one day he mysteriously disappeared in a storm. Legend says that the day he disappeared the sun became dark, day turned into night, and there were such furious winds and thunder that the people were terrified and fled. They later called that day the Flight of the People.

Many Romans believed that Romulus had been transformed into a god that mystic day.

FACTOIDS

Romulus once invited his neighbors, the Sabines, to a festival. During the festival, he abducted the Sabine women. The women eventually married their captors and later intervened so that the Sabines wouldn't attack the city in revenge. A lighthearted modern version of this incident is portrayed in the film *Seven Brides for Seven Brothers.*

Because a woodpecker helped the wolf take care of Romulus and Remus, it was considered a sin in ancient Rome to eat a woodpecker.

Rome was the first city in history to have a population of one million. London, England, achieved that number in 1810 and New York City in 1875.

DID YOU KNOW?

There are many stories of children being raised by animals, from Romulus and Remus to Tarzan. Most authorities believe these tales have no basis in truth and are simply legend or fiction. However, they admit that children isolated from all social contact can become "wild" children. One of the most famous is the "Wild Boy of Aveyron."

In 1799 an 11-year-old boy was discovered running wild and naked in a forest in France. He survived by begging for food from farmers and stealing from their gardens. After roaming the area for two years, he was caught and put under the care of Dr. Jean Itard.

Dr. Itard named the boy Victor and taught him how to dress himself and perform simple chores, but he could never teach him to speak. Although Victor could read and understand words to some extent, the only words he ever learned to say were *lait* ("milk") and *O Dieu* ("Oh, God"). The wild boy of Aveyron died at age 40.

Socially isolated "wild" children have also been discovered in modern times. In 1975, a social worker discovered a 13-year-old girl who had been kept isolated in a small room since about the age of two. Named Genie, she could not speak and had almost inhuman characteristics, including spitting, sniffing, and constant clawing.

Scientists studied Genie for five years and then put her into a series of foster homes where she was often abused, harassed, and punished. She eventually was placed in an adult foster care home where she received better care.

Although Genie acted like an animal at times, she was still human. Yet she wasn't always treated as a human or always as well as some people treat their pet animals.

Perhaps we still have a lot to learn about dealing with the "wild" children in our society.

What language is spoken the most in the world? (Hint: It's not Swahili.)

Mandarin Chinese, spoken by 885 million people, is the most prevalent language in the world. Spanish, spoken by 332 million people, is the next most popular. English is a close third, being the predominate language of 322 million people.

However, if all of the Chinese languages are included, such as Cantonese, Wu, Min Nan, Xiang, Jinyu, and so on, then there are over 1.2 billion people in the world whose primary language is some form of Chinese.

Chinese is spoken not only in China but also by ethnic Chinese in Brunei, Cambodia, Indonesia, Malaysia, Mongolia, the Philippines, Singapore, South Africa, Taiwan, and Thailand.

In Chinese, each character is pronounced as a single syllable, although words are usually two syllables. For instance, the character *ming* means "clear" and *bai* means "white." When the two characters are pronounced as one word, *mingbai,* it means "understand." Because there are no parts of speech in Chinese, most words can be nouns, verbs, adjectives, or adverbs, depending on their position in the sentence.

In spoken Chinese, a word may have four different tones. Although the word is phonetically the same, each of the four tones gives the word a different meaning.

Wouldn't it be great if English grammar were that simple?

FACTOIDS

There are 173 different languages spoken in California, including 23 different Native American languages.

More than 100 million Chinese have the surname Zhang. Over 40 percent of China's population has one of the 10 major surnames: Zhang, Wang, Li, Zhao, Chen, Yang, Wu, Liu, Huang, and Zhou.

In a language that uses an alphabet, the letters are clues to how a word is pronounced. However, Chinese characters are pictorial in form and give few clues to pronunciation. Thus, when reading a text, the reader pronounces the words according to the rules of his own language, or dialect. That is why Chinese speaking different dialects can write to one another and be understood. If they were speaking, they would probably not understand each other.

In San Francisco, California, the voter's ballot is printed in the three major languages: Chinese, Spanish, and English.

DID YOU KNOW?

Languages lead to literature and China has a long literary history, uninterrupted for over 3,000 years and dating back to the 14th century B.C. As in most cultures, China has many stories to illustrate some moral. Here is one of them.

An old man discovered that his horse had disappeared. Although his neighbors felt sorry for him, he said, "It doesn't matter. It may not be a bad thing, it might even be a good one."

One night the man's horse returned and had a beautiful companion horse with it. The man's neighbors were elated but he said, "Although I got a new horse for nothing, it may be good or it may be bad."

The old man's son loved the new horse and rode it often. One day he fell off the horse and hurt his leg so badly that he could never walk quite right again. The old man said, "Perhaps this will be a good thing."

A year later, the youth of the village were recruited to fight in a war. Because the old man's son couldn't walk quite right, he wasn't required to fight. All the other young men were killed but the old man's son was still alive.

This story tells us that good things can come out of bad. In adversity there are the seeds of new opportunity.

There is a great deal of wisdom in the ancient Chinese folk tales.

Where are the Spice Islands? (They're not on the shelf in your supermarket.)

The Spice Islands are actually the Indonesian islands called the Moluccas, and they include the famous island of Bali. They are north of Australia and south of Indonesia.

A source for cloves and nutmeg, the Moluccas became an important area for traders from approximately 300 B.C. Chinese,

Indian, and Arab merchants sought the rich profits from spices long before European traders came to the Spice Islands.

Because so many Arab sultans amassed great wealth by controlling the lucrative spice trade, one of the islands became known as the Land of Many Kings.

Europeans came to the Moluccas in search of cloves and nutmeg, which were prized as food preservatives. Wealthy women wore lockets filled with spices so they could freshen their breath easily, gentlemen added nutmeg to food and drink, and many spices were used for medicinal purposes to relieve colic, gout, and rheumatism.

The romantic name of the Spice Islands masked their long and bloody history of being ruled first by Arabs, then by Spain, England, and Holland. By the end of the 18th century, the spice trade had diminished and the Moluccas lost their importance. In 1949 they were incorporated into the Republic of Indonesia.

The Moluccas are mountainous and have frequent earthquakes and active volcanoes. The tropical climate produces up to 150 inches of rain a year. Because the Moluccas are a transition zone between Asian and Australian plant and animal life, many unique species of plant and animal life occur there. Over 20 percent of the bird species and 40 percent of the mammals are unique to the region surrounding the Moluccas.

FACTOIDS

Oregano wasn't a well-known spice in the United States until 1945, when American soldiers returned from World War II, bringing with them a taste for Italian pizza. In the 10 years following the war, the popularity of oregano increased by 5,200 percent.

A former official of the British East India Company, Elihu Yale, made a fortune that he later used to contribute to what was later named Yale University.

Saffron comes from a purple-flowered crocus. Three stigmas are hand-picked from the inside of each flower, spread on trays, and then dried over charcoal fires. A single pound of saffron represents 75,000 blossoms. It is the most expensive spice in the world.

Bay leaves come from the laurel tree. Ancient Greeks and Romans used wreaths of laurel to crown victors. The term "baccalaureate," which means laurel berry, comes from the ancient practice of honoring scholars and poets with garlands from the bay laurel tree.

Cinnamon comes from the bark of the cinnamon tree and is probably the most common baking spice. Romans believed it to be sacred, and Nero burned a year's supply of cinnamon at his wife's funeral. One of the motives for world exploration in the 15th and 16th centuries was to find new sources of cinnamon.

DID YOU KNOW?

The history of spices dates back at least to 3000 B.C. The earliest recorded use of spice is from an Assyrian myth claiming that the gods drank sesame wine the night before they created the world.

In about 400 B.C., the Greek physician Hippocrates listed over 400 medicines concocted from spices and herbs. About half of these are still used today.

Until A.D. 1200, the spice trade was controlled by the Roman Empire. Romans valued spices as highly as gold. When the Goths overran Rome, their leader demanded 30,000 pounds of peppercorns, along with gold, jewels, and silk to spare the population from death.

Later the Europeans explored passages to the East Indies in search of spices. Marco Polo helped establish Venice as an important trade port.

Christopher Columbus was searching for a direct western

route to the Spice Islands, or the Moluccas, when he landed in America.

Sweet spices have caused bitter wars. During the 15th through the 17th centuries, numerous wars broke out for control of the spice trade, the most notable being between the English and the Dutch.

The United States entered the spice trade in the 17th century. Between 1800 and 1890, nearly 1,000 United States ships made the around-the-world voyage to the Spice Islands. Later, Texas settlers developed chili powder and researchers in California developed techniques for drying onions and garlic.

Today the United States is the world's largest buyer of spices, followed by Germany, Japan, and France.

More questions? Try these Web sites.

Business and Social Customs Around the World

http://public.wsj.com/careers/resources/documents/cwc-countries.htm

Just click on the name of any country to read about its customs, including meeting people, body language, corporate culture, dining, entertainment, gifts, and helpful hints.

World Almanac

http://www.infoplease.com/world.html

This is your World Almanac on the Internet. It has facts about countries, disasters, architecture, geography, flags, and international relations. It has a number of other features including world statistics and current events.

Ocean Mysteries

http://floridasmart.com/subjects/ocean_mystery.htm

The ocean covers 75 percent of our planet and is virtually unexplored. It is the last frontier of our planet. This Web site discusses mysteries of the ocean such as the Bermuda Triangle and the Lost Continent of Atlantis. There is also an informative section on pirates.

The following two Web sites also have information about ocean mysteries, as well as other information about the oceans of the world:

http://montereybayaquarium.org/aa/aa_pressroom/pr_mysteries_deepfacts_pr.html

http://www.enviroliteracy.org/oceans_mysteries_of_the_deep.html

Countries That No Longer Exist

http://geography.about.com/education/geography/library/weekly/aa071999.htm

Have you ever wondered what happened to the countries of Siam, Bengal, Persia, and Rhodesia? This site will tell you. It's updated often, so the information is usually current.

World Facts Online

http://www.worldfactsonline.com/

This site lets you find information about countries throughout the world. You can select either global (worldwide) or national (select the country of your choice). If you select global, you'll see a lengthy list of links to other sites covering worldwide topics including capitals, composers, flags, maps, population, travel, weather, and much more.

If you select a specific country, you'll see one or more Web sites devoted to that specific country.

Exploring the Internet

Before discussing ways of finding information on the Internet, it's important to remember that anyone can put up a Web site and say anything they want to. In other words, a great deal of information on the Internet is false.

When you find an answer to your question, make sure you check its authenticity. For example, if it's the official site of the Discovery Channel or the National Football League, then the information is likely to be valid. If, however, it's someone's personal page, then it would be a good idea to double check with another source.

It's also important to know that even official sites sometimes disagree with one another. Even if you find what you need on an official site, it wouldn't hurt to double check with another official site.

If you don't want to use a general-purpose search engine, an excellent Web site that helps you find specific information quickly is at: http://websearch.about.com/internet/websearch/mbody.htm.

Click on "Web search how-tos" to see a number of helpful hints such as how to find a person's e-mail address, public records, driving directions, multimedia, free translation services, and much more.

Hundreds of Web sites allow you to find information on the Internet by searching on key words. I know of over 50 such sites, but rather than list them all, I'll give you just 7 that I've found to be most helpful.

ASK JEEVES
http://www.askjeeves.com/

FAST SEARCH
http://www.alltheweb.com/

ALTA VISTA
http://altavista.digital.com/

HOT BOT
http://www.hotbot.com/

YAHOO!
http://www.yahoo.com/

INFOSEEK
http://www.infoseek.com/

LYCOS
http://www.lycos.com/

To use a search tool effectively, you must try to be as precise as possible. For example, let's say you want to find a recipe for chocolate fudge cake.

If you search on "cake" you'll get a list of over half a million links (540,830 to be exact). If you are a little more specific and enter "chocolate cake" you'll see a list of over 13,000 links. If you next try "chocolate cake recipe" you've narrowed the list down to 222 links but that's still too many to look through. If you enter "chocolate fudge cake recipe" you'll see only four links. Of these, two simply have references to the cake while the other two are recipes.

If you run into trouble when looking for an item, try synonyms or related words. For instance, if you are looking for the "origin" of something and having difficulty finding what you want, try using the words "history," "beginning," and "start."

The search tools give explanations or tips on how to use them effectively. Be sure to read them carefully. For instance, when you use AltaVista you must put a + sign between the words.

If you search on "chocolate+fudge+cake+recipe" you will find only four links. If you search on "chocolate fudge cake recipe," the search tool will find every site that has the word *chocolate,* every site that has the word *fudge,* and so on. You'll end up with 656,740 links.

Finally, don't be surprised when you see what the search tool finds. When I was answering a question relating to jelly and jam, I searched on the word *jam.* In addition to the food, I discovered that rock groups share that name. There were many links to "Space Jam" and "Pearl Jam." There may be times when you search on a simple word and end up finding a rock group, a book, or a pornographic site. If that happens, refine your search and try again.

Comments Are Welcome

I'd love to hear from you if you have any comments, criticisms, or questions. You can e-mail me personally at:

Answerwhiz@aol.com

Index

About the Author

The Internet's legendary "Answer Whiz," **Bill McLain** was Xerox Corporation's official Webmaster. Responsible for the e-mails sent to the company Website, McLain and his team responded to an astounding 750–1,000 questions daily. While most of the e-mails he received were Xerox-related, every day scores of curious fact-seekers wrote with questions ranging from the bizarre to the useful to the downright comical. McLain collected the most memorable of these questions, along with his equally memorable answers, in this volume and in its predecessor, *Do Fish Drink Water?*